DOING PHYSICS

DOING PHYSICS

How Physicists Take Hold of the World

MARTIN H. KRIEGER

Second Edition

INDIANA UNIVERSITY PRESS *Bloomington & Indianapolis*

Portions of this book, reprinted herein with permission of the publishers, appeared in somewhat different form in: "The Physicist's Toolkit," *American Journal of Physics* 55 (1987): 1033–38; *Marginalism and Discontinuity: Tools for the Crafts of Knowledge and Decision* (New York: Russell Sage Foundation, 1989), chaps. 1 and 7; "The Elementary Structures of Particles," *Social Studies of Science* 17 (1987): 749–52; and "Temptations of Design," *Research in Philosophy and Technology* 10 (1990): 217–30.

This book is a publication of

INDIANA UNIVERSITY PRESS
601 North Morton Street
Bloomington, Indiana 47404-3797 USA

iupress.indiana.edu

Telephone orders 800-842-6796
Fax orders 812-855-7931

First edition published 1992.

♾ The paper used in this publication meets the minimum requirements of the American National Standard for Information Sciences–Permanence of Paper for Printed Library Materials, ANSI Z39.48–1992.

Manufactured in the United States of America

Library of Congress Cataloging-in-Publication Data

Krieger, Martin H., author.
Doing physics : how physicists take hold of the world / Martin H. Krieger. – Second edition.
pages cm
Includes bibliographical references and index.
ISBN 978-0-253-00607-3 (paperback : alk. paper) – ISBN 978-0-253-00608-0 (electronic bk) (print) 1. Physicists. 2. Physics – Methodology. 3. Physics – Philosophy. 4. Science – Social aspects. 5. Ethnology. I. Title.
QC29.K75 2012
530.092'2 – dc23

2012020453

1 2 3 4 5 18 17 16 15 14 13

FOR MY TEACHERS AND STUDENTS,
WITH GRATITUDE AND AFFECTION

Contents

Preface

Degrees of Freedom; A Note to the Reader; A Note for the Scholars; This Second Edition; Acknowledgments.

THIS IS A BOOK ABOUT HOW PHYSICISTS TAKE HOLD OF THE WORLD, actually about how some physicists get hold of some of the world. To an outsider watching physicists work, the details of that work and the physicist's obsessive concerns make little sense unless one has some idea what physicists are up to, what their various goals or purposes are. Technical moves *do* something, contributing to certain generic schemes. I want to describe the meanings of some of those moves, not so much to explain the physical world in some semi-technical or popular fashion, but to describe a rather familiar culture we all share.

For it turns out that physicists' goals have much in common with those of other theoretical endeavors which try to make sense of the world – whether by economists or anthropologists, for example – surely in part because those endeavors have been influenced by the work of physical science. And much of modern science developed in accord with economic and political modernization, the growth of both market economies and a strong sense of individual autonomy, and a spread of social alienation. The pervasive problem has been to find the right sort of individuals, and a culture in which such a liberal society might thrive. In vulgar terms, there is an identity of Cartesianism's particles and capitalism's actors and commodities. We might be said to have an economy of Nature.

Again and again, we shall see *analogies between physics and economics, political theory, anthropology, and sociology,* analogies that may be of interest to social scientists.

My claim here is that there is just one culture (rather than the two of C. P. Snow). For the culture of physical science is a subculture, articulating major themes of the larger culture – a larger culture whose ideas and practices have been, reciprocally, deeply influenced over the centuries by the physical sciences.[1]

A Note on Diction: I have deliberately used a number of colloquialisms, such as "getting hold of the world" or "getting a handle onto something," to capture the everyday experience we have in doing physics and to connect that experience with the larger culture. More generally, I have tried to use everyday terms to do technical work, the obligation being to use them consistently. When I describe physicists as being "obsessed" with certain models, I mean an insistent returning to a particular way of doing things and a recurrent compelling concern with certain issues, where such ways and issues might seem unreasonable to an outsider – in short, obsessions. In the same vein, I use "poignant" to describe the strange pervasiveness of physicists' commitments and, again to outsiders, the sometimes even sad doggedness with which these commitments are pursued.

Now, even if the technical moves physicists make are quite conventional and archetypal, the generic character of convention and archetype hides behind some concrete models and specific ways of going about things. Physicists will take the natural world as being much like the division of labor with its alienated individuals, or like a mechanism composed of parts, or like a system of exchange as in kinship, or like a black stage on which the drama can be natural phenomena. They get a handle onto the world by probing it, poking at it and seeing what happens. And, using the machinery of mathematics, they may analyze the meaning of common notions, and highlight and display various aspects of a phenomenon leading to a deeper understanding of the physics. They craft the world by using conceptual tools. Of course, such abstraction leaves lots out of consideration, and this is a good riddance, for it allows the physicist to get on with the work at hand. When physicists try to take hold of the world, to get a handle onto the world and shake that handle to see what will happen, they are quite willing to give up on most of the world so that what happens is simple and nicely related to their original shaking. They take hold of one "degree of freedom," and if they are lucky they have tamed the rest into silence.

James Clerk Maxwell, the great nineteenth-century physicist, put it nicely. He begins with a methodological remark and then presents a poignant clockworks-like mechanical analogy:

> We must remember that the co-ordinates of Thomson and Tait are not the mere scaffolding erected over space by Descartes, but the variables which determine the whole motion. We may picture them as so many independent driving-wheels of a machine which has as many degrees of freedom.
>
> We may regard this investigation [of ignorable coordinates] as a mathematical illustration of the scientific principle that in the study of any complex object, we must fix our attention on those elements of it which we are able to observe and to cause to vary, and ignore those which we can neither observe nor cause to vary.
>
> In an ordinary belfry, each bell has one rope which comes down through a hole in the floor to the bellringer's room. But suppose that each rope, instead of acting on one bell, contributes to the motion of many pieces of machinery, and that the motion of each piece is determined not by the motion of one rope alone, but by that of several, and suppose, further, that all of this machinery is silent and utterly unknown to the men at the ropes, who can only see as far as the holes in the floor above them.
>
> Supposing all this, what is the scientific duty of the men below? They have full command of the ropes, but of nothing else. They can give each rope any position and any velocity, and they can estimate its momentum by stopping all the ropes at once, and feeling what sort of tug each rope gives. If they take the trouble to ascertain how much work they have to do in order to drag the ropes down to a given set of positions, they have found the potential energy of the known co-ordinates. If they then find the tug on any one rope arising from a velocity equal to unity communicated to itself or to any other rope, they can express the kinetic energy in terms of the co-ordinates and velocities. These data are sufficient to determine the motion of every one of the ropes when it and all the others are acted on by any given forces. This is all that the men at the ropes can ever know. If the machinery above has more degrees of freedom than there are ropes, the co-ordinates which express these degrees of freedom must be ignored. There is no help for it.[2]

How physicists take the world is the way that world *is* for them – at least as physicists, at least for most physicists. If it is taken as a matter of the division of labor between particles and fields, that is just what it is. It is not *like* a division of labor, implying there might be a more authentic real existence. Rather, it *is* that model, as long as the model is productive. Surely, there are dis-analogies, leftover pieces, and misfits. Future, presumably better models may be very different from the current one, even while reincorporating its enduring insights. But all of this is always the case. Again, what matters is how productive is a model or a way of taking the world. If it is productive, the world *is* this way. Physicists may justify their taking the world in the ways they take it by means of an argument about its true nature. But in actual practice those justifications and references to its true nature are forgotten: The world *is* this way. In this vein, professional and craft practices generally treat the world as a given, suited to their models, whether it be in medicine or law or plumbing.

Again, I mean this book to give the reader a sense of what's up when physicists do their work: the moves, the rituals, the incantations. It is a cultural phenomenology, not a reductionist exposé. And it is not a textbook. There is no attempt to train the reader to do physics problems or to set up experiments. Nor do I work out the conventional technical formalism, or do derivations, or anything like that. Mathematics and formalism are wonderfully automatic in this field, like all such machinery when appropriately applied, doing all sorts of work by the way, that by-the-way work being physically interesting. (As we'll see the production people have to constantly attend to the machinery so that what appears automatic is in fact adjusted and repaired by hand, so that it can appear "automatic.") To have mastered the technical models, even in a freshman course, is to learn to become automatic in your practice: to think like a physicist, and presumably to be less aware of your conventions and archetypes *as* conventions and archetypes. Still, it would surely help to try out the various practices, even in toy arenas, whether it be by solving problems or by doing an experiment. Nothing is so hard to demonstrate than is the skill of noticing physically interesting phenomena. Laboratory courses usually are too programmed toward getting the right answer to allow the student to get really lost and waste lots of time. But what needs to be appreciated is just the possibility of there not being a right answer, of needing to fudge things by taking the world in one of the ways I describe, so you get someplace at all.

If this were a book in literature, it would be a book about archetypal themes and forms and structures, an "anatomy," to use Northrop Frye's term. Put differently, I want to display "the conventions and the craft" of doing physics.[3]

DEGREES OF FREEDOM

My plot is straightforward: to describe four dominant ways physicists conceive of the world, and to describe how they get at that world and find out about it, and the role mathematics plays in doing their work. In this enterprise: (1) there is a division of labor; (2) things are made up of other things; (3) everything that is not forbidden will happen; (4) whatever happens, happens on a stage; (5) we find out about the world by poking at it; and (6) using mathematical machinery we learn about the world by careful philosophical analysis of its notions and phenomenological description of its modes of appearance.* A handle onto the world is called a *degree of freedom,* whether it be the temperature of a gas, the position of a particle, or the orientation of a crystal. A degree of freedom is a direction for potential action, once we figure out how to take hold of that handle in an effective way. One needs to shake the handle with just the right energy, and in just the right direction, and one will hear the music of Nature in its purest tones. The harmonic oscillator, as in a pure tone produced by a tuning fork, is one of the prevailing models of good degrees of freedom – namely, its frequency and its size of oscillation or loudness. Now good handles often are deliberate setups – literally, set up – conceptually and experimentally. And if you choose the right ones the physics turns

*To give an abstract and technical epitome, perhaps best understood only after a first reading, the story goes something like this: The vacuum, when excited by a suitable input of energy, exhibits localized particles, whose kind and number is governed by a principle of "plenitude," the particles being seen by probes, and those particles may be composed to form larger entities.

P. W. Anderson describes two principles that will be pervasive in our discussion: ". . . broken symmetry, which tells us what the order parameter is and what symmetry it breaks . . . , and second, the continuity principle, which tells us to search for the right *simple* problem [of noninteracting particles] when confronted with a complicated one [of interacting particles] . . ." *Basic Notions of Condensed Matter Physics* (Menlo Park, Calif.: Benjamin/Cummings, 1984), p. 70.

out to be vastly easier to do and your understanding is manifestly more perspicuous than if you choose the wrong ones. (If you choose the wrong ones, as the physicist Steven Weinberg ruefully says, "you'll be sorry.")[4]

You want to arrange things so that the degrees of freedom you are talking about are the important ones. And hopefully, by your device and setup you have hidden or tamed the less important and potentially annoying degrees of freedom.

Note that I use *good* and *right,* more or less interchangeably, to describe degrees of freedom that lead to nice clean experiments and powerful and correct physical theories. The *good* degrees of freedom are ones that allow us simple handles onto a system (hence I speak of good handles); the *right* degrees of freedom lead to widely applicable theories. Of course, we want the good and the right to coincide.

Note, also, that I use *world* and *Nature* rather generically, and *system* as in "physical system" to describe an experimental setup or a conceptual abstraction in which "the physics" of the situation is highlighted (not the chemistry, not the politics). Finally, I use *story* rather than "explanation" to emphasize that the accounts physicists give are stage-setting and narrative, and rarely are they logical as such.

Experimentally, the physicist's goal goes something like: "This result shows that it is indeed possible, given enough filters and shields, to isolate a single degree of freedom in an object 'big enough to get one's grubby fingers on' from all other degrees of freedom sufficiently well to observe the quantum behavior of that degree of freedom."[5]

Before we begin, it will be useful to have some further examples of degrees of freedom. Again, a vibrating spring's degrees of freedom are its frequency of bounce and its maximum extension, or the position and velocity of a point on the spring. A drum's sound frequencies and overtones, and their loudness, its good degrees of freedom, are in effect determined by the shape and elasticity of its drumhead – and so you might hear the shape of a drum. A particle's degrees of freedom include its position, velocity, and charge. If the particle is not pointlike then the charge has a spatial distribution and that will have many ways of being arranged, and so there will be many more degrees of freedom. If the particle is a rigid body, then its spatial orientation and its modes of vibration are its degrees of freedom. If the rigid body could have a crystalline order, then the crystal's symmetries are degrees of freedom. And if the crystal is magnetizable, there are further degrees of freedom, its amount and direction of magnetization. And there are still hidden degrees of freedom, ones we do not see unless we heat the crystal, so that melting or chemical reactions can start taking place.

The degrees of freedom of a uniform gas or fluid include its temperature and pressure. If the gas or fluid is flowing and turbulent there are lots more degrees of freedom, for the pressure and density of the fluid will vary from point to point. If the gas were composed of different sorts of molecules, their relative concentrations would also be degrees of freedom. More generally, in systems having multiple components (for example, water and alcohol), the number of the various phases of matter (gasses, liquids, solids) that is allowed is a measure of the number of degrees of freedom of the system (such as temperature and pressure), namely, the Gibbs phase rule.[6]

Degrees of freedom are the ways a physical system might change or be different than it is just now.[7] And if we tie the system down in some way, its freedom is restricted and so are its degrees of freedom. Hence, notionally fixing the molecules of a solid in orderly crystalline places tames the degrees of freedom dramatically. Except, those molecules vibrate around those notionally fixed positions and hence there are now many vibratory degrees of freedom (unless the temperature is sufficiently low so that some vibratory degrees of freedom of the lattice must remain quiescent). A good handle onto a system is a degree of freedom that makes it possible to ignore lots of the others since they are otherwise constrained or held in place, and either dragged along with the degree of freedom or left untouched by it. The temperature, for example, is often a very good handle since it determines the extent of excitation of all the vibrational degrees of freedom of a solid in equilibrium. (A bit of detail: The excitation of a degree of freedom requires a quantum of energy, one that is quite rarely available if the temperature is low enough compared to the quantum size (which is proportional to temperature). This fact is used to explain dilemmas in the classical account of specific heats, namely the hiddenness of the degrees of freedom of the core electrons in a solid. For those electronic modes, at about one electron-volt of energy, are not excited at room temperature, equivalent to a few hundredths of an electron-volt average energy, and so they do not contribute to the specific heat.)

More generally, if there are no degrees of freedom then the world is fully necessary. And so there are accounts of creation that allow for no free variables. And if there is an infinitude of degrees of freedom, where none of them is constrained, nothing fixing things in place, then the world is fully arbitrary. The actual world, as physicists deal with it, is somewhere in between; and I want to sketch how physicists make their peace with that somewhere in between.

In sum, my purpose here is to describe the ways physicists are convinced that the world *must* go, their tradition of models and techniques and phenomena that delimit for the most part what they take as Nature. Here I have in mind an often heard phrase, say, concerning a yet undefined physical situation or problem: "it must go this way" – immediately leading to a suggestion for a simple model or an emendation or a speculation. Here *must* is a combination of reasonable guess, skillful craft-work, and a sense of Nature's character. One would be genuinely surprised if Nature did not go this way.

I want to retell and interpret the stories physicists tell when they take hold of the world. Of course, all of this is an "of course" to a physicist – or at least I hope so. But it is not so obvious to outsiders; nor are the cultural connections, as conceived explicitly, so much part of being a well-trained physicist. Still, again, I would hope that for the practicing physicist my description would possess the ring of truth (to use physicist Philip Morrison's term), leading to a greater integration of what the physicist already knows and to a moment of self-recognition.[8]

ISING MATTER

In chapter 6 I ask, How does mathematics do its work in physics? What is the structure of argument in mathematical physics? My main point is that mathematics is machinery or a tool for doing physics; and, it is a form of philosophical analysis and of phenomenological description. The technical demands of rigor and precision are not merely for show. They reveal more of the physics of the system being described and analyzed. I

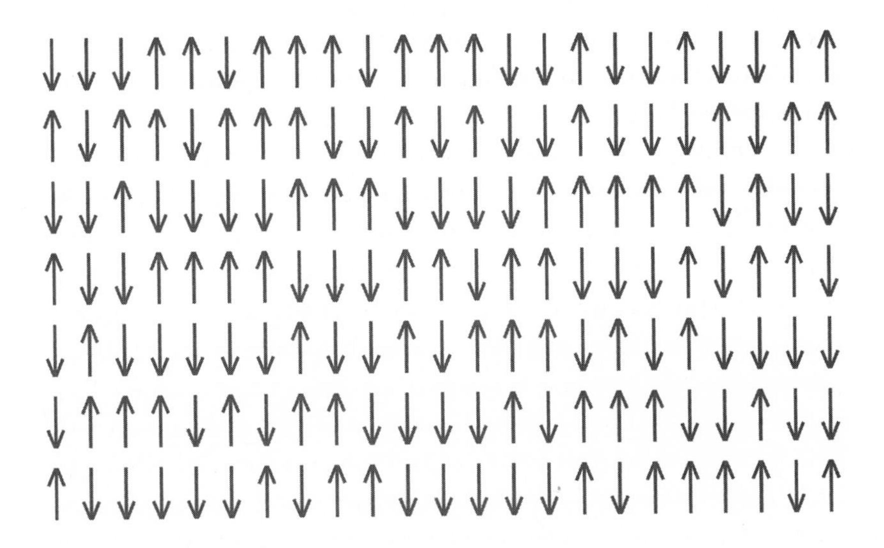

P.1. The Ising lattice in two dimensions at a high temperature

use many of the same examples as earlier in the book, namely the mathematical modeling of ordinary bulk matter composed of molecules, and the mathematical modeling of a phase transition such as liquid freezing or an iron bar becoming permanently magnetizable – where by "mathematical modeling" I mean expressing a physical system in mathematical terms, the word "modeling" implying that the expression is schematic and incomplete. As preparation, it may be useful to say a bit more about one of these models, the Ising model of ferromagnetism. The model appears in chapter 4, describing a phase transition as a matter of scaling and choosing the right degrees of freedom, and in chapter 6 as an example of a mathematical *tour de force*.

Schematically, one pictures a piece of iron as a two-dimensional grid or lattice of atomic magnets, each of which can point up or down.[9] The atomic magnets bounce randomly, each on its own, rapidly oscillating from up to down and back, due to "thermal motion" (much as air's molecules move rapidly at room temperature and pressure, bumping into each other many times each second as well as bumping into walls of the enclosure – namely, the pressure). Yet there is also a magnetic force among pairs of adjacent atomic magnets that aligns them with each other. There is a conflict between disorderly effectively-random thermal motion and the ordering force of magnetic alignment. If the temperature is low enough the magnetic alignment force dominates; in fact, that transition to dominance occurs at a well-defined "critical" temperature. Let us call this model of matter "Ising matter," after the author of the earliest papers that described its behavior.

The mathematical problem of solving this model, going from the atomic situation to ordinary everyday bulk matter, and determining that critical temperature, was solved by Lars Onsager in 1944, and in the subsequent years there have appeared many different mathematical ways of solving the problem. Some just literally count up all the interactions among the atoms: one by one, or in blocks of spins of increasingly larger units. Some discern regularities in the lattice system and the magnetic-thermal forces – such as that a very disorderly high temperature system with a bit of order is like a very orderly low temperature system with a bit of disorder; or, the system looks the same at all scales, so if you get closer you see the same patterns; or, that scaling would seem to define the algebra of devices used to do the counting-up. Some find "particles" (actually orderly rows of spins) in this lattice and work with them. Some model that lattice as a field. All these points of view are it seems true; Ising matter accommodates them all, and we might say that there is an identity in that manifold presentation of its profiles. All the methods come up with the same answers (as we might hope), and a retrospective reading of Onsager's paper suggests how all these methods are built into his solution – although that is apparent only retrospectively. How and why the very different mathematical technologies or methods are applicable is not always easy to discern. It would appear that we have an analogy among these methods, and then an analogy of this analogy with a similar analogy in pure mathematics.

Not only is Analogy Destiny; it would seem to be Analogies all the way Down.

A NOTE TO THE READER

Some of this book is hard going. So I should perhaps say something even more explicit about audience, difficulty, and ethnographic distance, so that the reader will have appropriate expectations for a book that at first might seem to be a popularization of physical science when it is actually an account of aspects of a subculture in our society, a description of the world as physicists take it.

I have tried to write so that readers who are not physicists will readily follow most of the text, employing their everyday intuitions to understand an arcane subculture within their own society. What will help, of course, is that it is a *sub*culture, one sharing in the general culture's central themes and rhetorics. The reader must have some experience of the general culture, say of a factory as a division of labor, so that the models I describe are seen as models. Otherwise, the culture to which I am referring would be as obscure as the physicist's subculture.

The problems with this approach are twofold: First, again, many readers will think of themselves as laypersons; and so they might well expect a popularization, an explanation of the physics. And what they receive is an account of a culture and a rhetoric, about which they are as expert as anyone. On the other hand, for physicists the technical material is more or less obvious. But the cultural and metaphoric account will seem suspicious, since it shifts their everyday work into an alien context. I have tried to put sufficient technical explanatory material in the notes to take care of the arguments I would want to make to these native specialists, especially concerning fine points. I would also hope that physics students might find the stories I tell illuminating, helping them to have richer intuitions about what is "really" going on in their technical courses.

(Technically, I have taken a very particular point of view on physics, much influenced by contemporary ideas in quantum field theory of many-body systems. I imagine that another point of view would produce a different set of models and modes of getting at the world. In any case, I have not at all emphasized the currently popular "mysteries" of modern quantum mechanics, staying within rather more orthodox interpretations. I have made much use of some comments by the very unmysterious physicists Steven Weinberg, P. W. Anderson, and Richard Feynman.[10] The seminal ideas of John Wheeler and Lev Landau are crucial, especially for chapters 1 and 4. What is impressive to me is how the traditional issues and metaphors of mechanical philosophy are replayed in new contexts.)

That some passages are unavoidably "hard" reflects our distance from this seemingly arcane subculture, not that science is difficult per se. However, when I use the term "technically" (as I did in beginning the last paragraph) I am setting a warning flag, indicating that a passage is for the specialist. I set that flag sparingly in the main text, but I have been less restrained in the notes. In rereading the notes, I discovered that I had often assumed that the reader knew the conventional meaning of letters and symbols (for example, ω = frequency). For this is a highly conventionalized culture, even if the conventions may well change in time and location. In any case, I have endeavored to fill in the definitional lacunae.

Chapter 1 sets in place much of the material needed for the rest of the book. The reader is welcome to read the initial paragraphs of each section, returning to the detailed mechanisms later on.

Again, this is an ethnographic or cultural report on the technical practices of a subculture. When I say that "physicists believe" I mean that a quite recognizable, not at all idiosyncratic group of people think this way – but not all physicists think this way. And, again, my purpose here is to give the reader a feel for what it is those physicists are up to and what that has to do with their cosmos. The payoff, both for layperson and for physicist, is to see the work of science as sharing in the work of the society.

As for voice, I have shifted from referring to physicists in the third person to putting ourselves in their place. One of my goals, and perhaps that of much of cultural ethnography, is to show how *we* could be one of *them* (and of course, some of my readers are physicists).

A NOTE FOR THE SCHOLARS

Here I hazard some brief comments on whatever import this kind of description might have for conventional studies of science by philosophers and sociologists and historians.

My main claim here is to provide an analytic description of some of the work of physicists, one that they would find recognizable. Again, it should possess the ring of truth. Almost all descriptions of science by social scientists and historians and philosophers are seen by practicing scientists as strange or as missing the point or as demeaningly ironic. This does not mean that these latter descriptions are wrong, but rather that they are governed by the demands of scholarly inquiry within particular disciplines.

My second claim is that a description such as the one I provide justifies science (to use theological terminology) in the sense that it places science within the larger culture. Students taking broad courses in general education (Contemporary Civilization or Humanities, as I did) should not be surprised by the ideas they find here. Such a description is adequate not because it is well argued, with the implication that argument leads to conviction, but rather because there is sufficient detail, provided both in the scientific and in the cultural arenas, that recognition is transformative – we think of something in a new way. Of course, argument about details is in the end crucial. For scholarship is, by its traditional definition, scholia or commentary. But here I want to set forth a thesis in its broadest outline.

I think this kind of description is fundamental, in that it provides the material which makes possible philosophizing or sociological theorizing. Of course there is a good deal of prejudice about such issues built into the description. I am in effect anti-foundationalist; I am insisting on the practice of physics as it is done. And I am in effect against the ironic tone of much of constructivist analysis, for I believe that the analogical structures are necessary: namely, if we are to do physics as physicists understand that endeavor, it more or less must go one of these ways. For these analogies are transcendental, the grammar of physics. Moreover, there is a rather small number of ways – hence my talk of tools, and of economy, mechanism, kinship, and theater. The actual repertoire is just that, a repertoire not an infinity.

I am claiming as well that physics is subject to a cultural analysis. Its technical features, no matter how mathematical or mechanistic, are subject to the kinds of interpretation performed by critics of literature and art and by archaeologists and anthropologists. Those features are encrusted with meaning. When I speak of a rhetoric of Nature, or of economy or of kinship, I mean that science is subject to the same kinds of discourse that other human activities are subject to, whatever claims to truth each may make.

When I read the philosophic literature, I am most comfortable with the work of Thomas Kuhn and Ian Hacking.[11] Kuhn strikes me as being very close to what the physics is really like, and his notion of paradigmatic exemplar covers much of what I mean by analogy and by concrete archetypal example. Hacking's emphasis on "intervening" is just what I mean by handles, both experimentally and theoretically. I am less sure where I stand on many of the traditional philosophic issues, say as Hacking describes them in the "representing" half of his book. But rather than asking what the world is really like, I would rather say how we take hold of it and so describe its phenomenology.

The analogies which concern me here are cultural analogies, stories or narratives connected to other such stories, with no necessary mathematical or structural link.[12] It is in the terms of art and how they are used and what they refer to, or in the technical tasks and how they are carried out and the other tasks they are linked to, that the analogy is made apparent. I have given a great deal of discussion of model and analogy under the rubric of tools and toolkits in a previous book, *Marginalism and Discontinuity: Tools for the Crafts of Knowledge and Decision* (1989), and will not repeat it here. When we talk about tools, what is crucial is that tools are used to do work. A set of tools provides a provisional way of taking hold of the world and doing something with it. Toolkits have a small number of tools and we adapt those tools to new situations. Hence the small number of major analogies I use here.

Social studies of science have shown that "practice should be seen as a process of modeling, of the creative extension of existing cultural elements."[13] Such extension is contingent and open-ended, the exact extension of a model dependent on how it is taken to fit a new situation. Good models have a high degree of analogy with what they are to model, along the way requiring modification if they are to overcome initial mismatches. Put differently, insofar as physicists are Kantians with no direct access to Nature, they are committed to allegory and imagery – much as the pastoral theologian, such as Augustine, employs allegory for lack of direct knowledge of God (as a consequence of the Fall).[14] The physicist's commitment is expressed not so much by a creedal statement, but by the presumption that the world *is* this way, the world *is* this allegory.

One might ask how I decided which are the major analogies or models. Some, of course, are venerated in myth and scholarship – such as the clockworks. Others play such central roles in our culture, such as economy and kinship and craftwork, that we are not so surprised to see them repeated in a subculture. And others, such as the theatrical stage, are happy realizations that remind one that science is much like the arts in that it is an orderly provision of the world. Other major analogies, such as that of evolution and organism, seem to play a much smaller role in most of physics. In the end, I think one justifies a cultural analysis by its value in epitomizing a wide variety of phenomena, its recognizability to its practitioners, and its being a repetition of analyses for other aspects of the culture.

I do want to emphasize that whatever Nature does, Nature does its work not verbally or textually but through physical interactions. That the everyday phenomenology and the physics go together is perhaps not ultimately surprising; but, to me, how that "going together" takes place is, as craftwork, wondrous and remarkable.

Finally, a brief remark concerning history of science. What I have tried to do here draws from the history of ideas and culture and science, in that it insists that contemporary notions have a history, a history of repetition and modification of previous notions. Just how self-conscious scientists are of economies, mechanisms, kinship and plenitude, stages, and toolkits is a matter for historical scholarship. For that consciousness surely changes, some larger cultural notions going into comparative eclipse for a while. Moreover, such a history of science is not reducible to a history of ideas or of economic relations. Scientific events – experiments and phenomena – will resist ideas and economies, a resistance that then leads to real work for the scientist.

THIS SECOND EDITION

Rereading the book so many years after it was first published has been a curious experience. Almost on every page I would think of something I left out or an apparent error, or that some proviso or modification was needed. I would check the notes, and discover often that I had dealt with the issue. Or, I found that perhaps two pages hence in the main text there was the needed discussion.[15] And, there were other errors, conceptual, technical, and verbal, that I have corrected. (Surely, others remain.) I was repeatedly struck by my commitment to the themes of Analogy is Destiny and to The Craft of Doing Physics, and again how my work on an earlier book, *Marginalism and Discontinuity* (1989), is a foundation for *Doing Physics*. In the more than twenty years since I wrote *Doing Physics*, I have written two fairly technical books on how specific mathematics and models realize those analogies and enable that craft: *Constitutions of Matter: Mathematically Modeling the Most Everyday of Ordinary Phenomena* (1996) and *Doing Mathematics: Convention, Subject, Calculation, Analogy* (2003). For this edition, in chapter 6 I have provided a nontechnical epitome of those two books, while making minor changes throughout the original text and notes.

In some of my other work, as a professor of city planning, I have spent a good deal of time in factories and workshops in Los Angeles. I realized that I was following in the footsteps of the encyclopedist Denis Diderot, who with d'Alembert are the authors of the *Encyclopédie* (1750–1772). Diderot tried to describe and illustrate the *arts et métiers*, the crafts and manufacture of his time, the actual practices of the workers. I, too, have been describing some of the crafts and modes of manufacture of physics: the design of a factory, the engineering design that produces an object out of components, and so forth – the actual practices of the workers, the physicists. I have focused on the conceptual and theoretical work, not on the design of experimental setups. And for the most part I have focused on descriptions that are microscopic and molecular physical processes, rather than the macroscopic (as in celestial mechanics or the proverbial block-and-tackle pulley).

One last proviso. This is not a book describing the practical how's of doing physics, even theoretical work. For example, here is a description of the ways of working of one theoretical physicist, John Bardeen:[16]

- Focus first on the experimental results via reading and personal contact.
- Develop a phenomenological description that ties different experimental results together.
- Explore alternative physical pictures and mathematical descriptions without becoming wedded to any particular one.

- Thermodynamic and other macroscopic arguments have precedence over microscopic calculations.
- Focus on physical understanding, not mathematical elegance, and use the simplest possible mathematical description.
- Keep up with new developments in theoretical techniques – for one of these may prove useful.
- Decide on a model Hamiltonian or wave-function as the penultimate, not the first, step toward a solution.
- DON'T GIVE UP: Stay with the problem until it is solved.

ACKNOWLEDGMENTS

The research for this book was supported, both at the Massachusetts Institute of Technology and the University of Southern California, by grants from the Exxon Education Foundation. The support of Robert Payton and Arnold Shore, then at Exxon, was very important. I am grateful to my colleagues at MIT, especially Larry Bucciarelli, Carl Kaysen, Evelyn Keller, Michael Piore, Roe Smith, Sharon Traweek, Leon Trilling, Sherry Turkle, and Charles Weiner. The final revisions were done while I was the Zell-Lurie Fellow in the Teaching of Entrepreneurship at the University of Michigan.

My teachers at Columbia taught me how to think like a physicist. Those teachers thought it important to speak to "the other side of campus," as my advisor Leon Lederman puts it. I still do. Almost twenty (now forty) years ago I was a fellow at the Center for Advanced Study in the Behavioral Sciences the year that Yehuda Elkana, Robert Merton, Árpád Szabó, Arnold Thackray, and Harriet Zuckerman were there – and to boot, Chie Nakane and Terry Turner were also fellows. They made it possible for me to be fruitfully struck by the fact that the revolution in particle physics in the 1970s (the "standard model") was once more a repetition of the structures (namely, Maxwell's equations) we had seen before, much as my teacher of classical mechanics, Herbert Goldstein, insisted on quantum mechanics' being a repetition of classical mechanics, suitably understood. I owe to Hunter Dupree, at the National Humanities Center, the conviction that all of this is about science.

At the University of Southern California, Paul Bohannon, Alan Kreditor, and Karen Segal gave me the chance to teach an honors science course for nonscientists, from which this book arose. David Richardson gave an early draft a close reading. Abraham Polonsky has been the kind of literate fan – recognizing just what you are up to – one wants when writing a book or a screenplay or even in getting through life.

My friends have taught me a very great deal, and besides the many persons mentioned above let me add Jay Caplan, Tom and Jehane Kuhn, Eric Livingston, Andy Pickering, Gian Carlo Rota, Sam Schweber, and Gerry Segal. And Miriam Brien, Susan Krieger, and Elizabeth Kuhn. And John Bennett. And there are more.

No parent writes a book without a child who goes to sleep on time. For that, and a lot lot more, I love you David.

As for this second edition, many of the colleagues and friends mentioned above have passed away. I will not repeat the acknowledgments in my later books *Constitutions of Matter* and *Doing Physics,* for perhaps not surprisingly, they are much like what I have written above. Bob Sloan of Indiana University Press encouraged this second edition. And my work in this area, while not supported directly, benefited from a variety of foundation grants.

My son, David, is now a young adult still asking the best of questions.

DOING PHYSICS

ONE

The Division of Labor: The Factory

Nature as a Factory; Handles and Stories. What Everyday Walls Must Do; Walls for a Factory; Walls as Providential. Particles, Objects, and Workers; What Particles Must Be Like; Intuitions of Walls and Particles. What Fields Must Be Like.

THE ARGUMENT IS: THE WORKINGS OF NATURE ARE ANALOGIZED to a factory with its division of labor. But here the laborers are of three sorts: walls, particles, and fields. Walls are in effect the possibility of shielding and separation; particles are the possibility of sources and localization; and fields allow for conservation laws and path dependence. Different kinds of degrees of freedom are associated with each type of laborer, and the laborers naturally restrict each other's degrees of freedom – if the Factory of Nature is to be as productive as it is. Corresponding to the efficiency of the division of labor in a factory or an economy is the comparative richness, elegance, economy, and wide applicability of a physical mechanism or theory or model. Technically, Maxwell's equations for electromagnetism are one realization of this political economy of a transcendental aesthetic, to honor both Adam Smith and Immanuel Kant in one phrase.[1] (We discuss other mechanisms of production in subsequent chapters, for example ones in which exchange and the extent of the market are crucial features.) My claim is that physicists take Nature in this sense of manufacture; of course that sense being interpreted in terms of empirical "peculiarities," as Smith employs the term.

I

NATURE AS A FACTORY

Here is Adam Smith in the beginning of *The Wealth of Nations* (1776), describing the division of labor:

> The greatest improvement in the productive powers of labor, and the greater part of the skill, dexterity, and judgment with which it is any where directed, or applied, seem to have been the effects of the division of labor. . . .
>
> But in the way in which this business [of pin making] is now carried on, not only the whole work is a peculiar trade, but it is divided into a number of branches, of which the greater part are likewise peculiar trades. One man draws out the wire, another straights it, a third cuts it, a fourth points it, a fifth grinds it at the top for receiving the head; to make the head requires two or three distinct operations; to put it on, is a peculiar business, to whiten the pins is another; it is even a trade by itself to put them into paper; and the important business of making a pin is, in this manner, divided into about eighteen distinct operations, which, in some manufactories, are all performed by distinct hands, though in others the same man will sometimes perform two or three of them. . . .
>
> This division of labour, from which so many advantages are derived, is not originally the effect of any human wisdom, which foresees and intends that general opulence to which it gives occasion. It is the necessary, though very slow and gradual, consequence of a certain propensity in human nature which has in view such extensive utility; the propensity to truck, barter, and exchange one thing for another. . . .
>
> As it is the power of exchanging that gives occasion to the division of labour, so the extent of this division must always be limited by the extent of that power, or, in other words, by the extent of the market. (Book 1, chapters 1–3)[2]

The great invention here was to appreciate that in order to make pins or anything else, and to understand how they are made, one divides the work into specialized functions (those "peculiar trades"), attributes those abstracted functions to individual workers, and then provides for a system in which their labor is coordinated. Such an economy or a factory turns out to be both efficient and comprehensible. No individuals need do everything for their own livelihood, as they might on a farm. Nor would they need do everything to make a piece of equipment. What is needed is a mechanism to make sure that each individual knows what to do, and a means of organization and communication – whether it be a factory with its distinct tasks and processing lines, or a market economy with its specialized jobs, processes of exchange, and the prices attributed to labor and to goods. Such a division is not only efficient, it readily allows us to pinpoint what is going wrong if the factory does not function as we expect it to: some specialized task is not being done properly, or some particular means of coordination has become sticky. Rarely, if ever, is the whole factory to be reorganized. One almost always need merely to get hold of some specialized part and fix it.[3]

Of course, it is a very great achievement to create such a factory or economy, to figure out a workable division of labor and a mechanism of production. Careful prior analysis may help, but often it is a matter of trial and error, and perhaps even of settling into a configuration that is not the best one, but at least it works – as David Hume (1779) would have suggested, a consequence of its having been until then "botched and bungled."[4]

Now imagine that we, as economic anthropologists, were to come upon a seemingly productive system, and then tried to figure out how it worked. We may have some general ideas about how factories are organized and have some particular models or examples in mind. If the system just fits our ideas and templates, we are, so to speak, in business. But this particular system may be of a different shape and size, its boundaries uncertain or idiosyncratic. It is not quite so manifestly analogous to our models, not quite so readily gotten hold of with our regular toolkit – or so it seems. So we try out a tentative organization-and-flow chart drawn from our ideas, models, and tools, and see if it makes any sense of the workings of the factory. Along the way, we have to label the workers and work stations correctly, the product has to be distinguished from the garbage, the sections of the factory have to be delineated. Eventually, we might begin to understand how the factory works, why it is productive, and how it might break down and so exhibit new phenomena, and what to do to repair it if it does break down. (Recently, I have had this experience in an actual workshop, a small foundry.)

Such is the task, I would argue, that many physicists see themselves as taking on (as do many a theorist more generally) when encountering the world. Nature is in effect taken to be a factory or an economy.[5] Can the physicist discern a division of labor within Nature, and a mode of organization, that makes sense of what Nature is doing – in that sense of a factory?

Soon after Smith, Immanuel Kant too provided a way of thinking of the division of labor required to make up Nature as physicists came to view it. The Transcendental Aesthetic that begins *The Critique of Pure Reason* (1781) might be taken as suggesting that space is just what is needed, grammatically and physically (what Kant called the "transcendental condition"), for objects to be separated and distinct from each other, and that time is the condition for there to be sequences of events and a causal relationship among them. Here, the natural division of labor in making up the world is between objects and space, between events and time. So we might ask: Which properties do we give to discrete objects, which to field-like space, and what mechanism do we prescribe for their interaction, so that we have an account of how the world works?[6]

I take it that the physicist's initial problem is to discern "the political economy of the transcendental aesthetic": (1) to describe the precise modes or mechanisms by which objects are delineated and so separated from each other – the *walls,* shields, and surfaces; (2) the names or labels or properties through which objects have their own identity and are influential in the world – *particles;* and, (3) the provision and delineation of space with its own properties, so that in space's interaction with particles we have an account of Nature's workings – *fields.* As in a factory, the various laborers work together to produce Nature, according to rules which are often traditional and conventional – such as the rules that interaction between particles is "local" rather than "at a distance" and that neither particles nor fields have a memory of their past. Other divisions and rules are possible, but if the factory is to be productive the divisions and rules have to work together.

In my discussion, walls, particles, and fields are all taken to be laborers.[7] Now, we might think it more natural to treat particles as most directly analogous to workers, and walls (and perhaps fields) as material and capital infrastructures much like the factory building and its machinery. But here I treat labor and capital as qualitatively similar inputs, so to speak, much as do economists in their formal production functions. I want to describe how they work together, deliberately avoiding any argument about particles vs. fields. As for the factory building (the mechanisms of interaction), we shall discuss its organization later in this chapter and in subsequent chapters.

In chapter 2 we describe the various kinds of individuals suitable for a factory or for an economy of Nature; in chapter 3 we delineate how exchange and the extent of the market define the factory; in chapter 4 we show how we set up both a factory and its outside suppliers so that the factory's production process is fairly straightforward; in chapter 5 we describe how an industrial engineer would investigate the factory's workings and the toolkit needed for making sense of such a factory; and in chapter 6, we describe some of the mathematical machinery in that factory and how scientists creatively use that machinery to do some of the work of physics.

Our first problem will be to describe the dynamics of the walls or shields, how things are kept apart or separated from each other so there could be space between them. Once we appreciate how walls are designed, then the design of particles and of fields follows in a natural way. But before trying to describe walls in some detail, I want to say a bit more about the task we are up to.

HANDLES AND STORIES

The attempt to make sense of Nature in terms of a division of labor may be thought of as participating in one of the abiding human endeavors: an attempt to articulate and analyze our experiences and the phenomena we encounter, in order to provide ourselves with a handle onto the world. Put differently, if we can manipulate the world we can understand it. Now the handles that will concern us here are the degrees of freedom – for example, position, temperature, charge, pressure, energy – of systems physicists concern themselves with. (The Preface provides an introductory discussion of the notion of degrees of freedom.) And those handles or degrees of freedom may be seen to be characteristic features of the laborers (walls, particles, fields) that make up the factory that produces Nature.

Our task here is to describe how physicists go about finding handles, setting up situations which are so to speak handleable, and how they view those handles as part of a coherent story or a theory of both manipulation and understanding. Such a description is perhaps much like the anthropological ethnographer's: What do these people (here, physicists and their community) do in their conceptual and practical work, what is the meaning for them of what they do, how does it make sense in their terms and in ours, and what kind of world or cosmology does it provide? As we shall see, what is striking in this kind of description is the obsessiveness and poignancy of people's commitments to their ways of going about their work. (On the use of such terms as obsessive and poignant, see the Preface.) No matter how difficult and peculiar it may seem to outsiders or even to themselves, their commitment to the practices and strategies is practically total. And the work, no matter how technical and purportedly arcane, may be seen in terms of tasks and motives we more generally share. Now, I should note that I am not talking here about the actual division of labor among scientists and others in the doing of science, about its social and bureaucratic character. Here the division of labor is that of Nature, namely, the divisions employed in physicists' conceptualizations.[8]

Again, in this chapter we shall see how physicists are interpreting Nature as a factory; in the next, as a collection of parts that fit together; in the third, as a system of interrelationship and interaction and exchange, much like Smith's economy or in kinship and marriage; in the fourth, as a theatrical stage that displays an abstracted if everyday world, the motivating problem being how something arises from nothing; in the fifth chapter, how physicists interpret Nature as something to be handled and poked and so observed and changed; and, in the sixth chapter we describe how mathematics provides a supple language and a machinery physicists create and employ for modeling and understanding Nature. In sum, these physicists go about inventing a division of labor for Nature, one in which things are made up of other things, in which everything that is not forbidden is allowed, where whatever happens

happens on an empty stage, where we find out about the world by poking at it, and we learn to talk about the world, in a variety of dialects, using mathematical machinery.

II

WHAT EVERYDAY WALLS MUST DO

A wall creates two sides, an inside and an outside, a left side and a right side, a core and a periphery, a black box and an external world, a body and an environment. Walls divide the world into separate rooms, or discrete particles, or isolated and enclosed and demographically-addable individuals – perhaps with space between them.[9] Now, much of physical science is a story of individuals in interaction, whether it be interactions of atoms in chemistry or of elementary particles in physics. Physicists often then take as their task the creation (or invention or discovery, as you will) of just the right kind of walls, with just the right possibility for being breached, so that there may be the right kind of interacting individuals so as to manufacture Nature. So, walls in thermodynamics may define suitably restricted systems which then, say, have definite temperatures, those walls perhaps breached by heat or material. Or, the valence cloud of an atom's outer electrons, participating in the chemical bond and so acting as a wall, in effect hides the chemically uninteresting features of an atom or molecule – they are too tightly bound to interact – and yet allows for a breach of energy or charge at the meeting place of the atom and the world of other atoms around it.

Now there is no single canonical definition of what a wall must be like. Rather, there are a variety of archetypal cases and conventional analogies that instantiate what a wall must do and just how it does that walling-off. Some of these notions of walls will seem peculiar and strange. But it is just that strangeness we feel that tells us that this is a conceptual and practical invention, deriving from a set of experiences and necessities we ourselves may not have had – but could have. I want to look at the kinds of walls – the kinds of conceptions of walls – that physicists need to do their work, to analyze the production of Nature. ("What kinds of walls does Nature need?" is perhaps an allowable concision.) For purposes of exposition I place ourselves ("we") in the role of a physicist.*

Everyday walls may be defined as boundaries, interfaces, functions, skins, and dynamical processes. Boundaries delineate separation, interfaces describe permeability and interdigitation, functions allow for specific conditions to be maintained at the wall, skins hold together and bind, and dynamical processes respond to the outside world.

The wall may be a *boundary* line, like that between nations. Such a boundary might also allow for interchanges of specific goods in specific directions, and it might maintain certain conditions on itself (of purity or of temperature, for example). The boundary line and its conditions would seem to have to be maintained actively, by border guards, so to speak, if the boundary is not to fall apart. Yet, still, for many analytic purposes we need merely specify the spatial separation that the boundary defines (or its topology) and its exact shape.

Now, that boundary may be between two fluids which do not ordinarily mix, an *interface,* say between oil and water. Interfaces are breached by processes of mixing and intermingling and interdigitation. We might add soap to the water, or in the case of a water-ice interface we begin to melt the ice. The area of the interface can become very large, with fingers of one material jutting out into the other, just what we might mean by intermingling and interdigitation.[10] In effect, the interface has become a permeable wall, allowing material from each side to enter the other.

As I have indicated, some walls are conceived of in *functional* terms. They *do* something. They hold temperature or electric charge constant, or prevent heat from escaping, or ensure that interactions with the rest of the world are weak – by some means. When these interactions are weak,

*Guidance to the Reader: In order to capture the texture of these endeavors, the material in this part is comprehensive and detailed, and in that sense it is technical. My presentation is quite closely motivated by the structure of contemporary physical theory as well as the phenomenology of physics. Again, the purpose here is to appreciate the moves physicists make, and thereby appreciate what they mean by Nature. And what I have tried to do is to systematically catalog those moves. Some readers may just want to scan the pages in which the detailed phenomenology of what walls do is presented.

the enclosed objects can be more independent of each other.[11] Ordinarily, we do not inquire, at least in theoretical and conceptual discussion, about the size or nature of such a wall, or just how it works. We are concerned with its functionality.

In contrast, consider a binding *skin,* such as a balloon, or as on an apple, or the surface of a solid ball or a nucleus. Surely these walls are functional, but we are acutely aware of their thickness and composition and resilience, and more generally that they have to protect, face the outside, and perhaps hold in something. And, dynamically, stuff could vaporize off such a wall, or accrete onto it. Thinking of a balloon, we expect that the skin balances the inside and outside forces; thinking of a liquid's surface or of a nucleus, the skin balances what might be vaporized off of it with what might be condensed onto it.[12]

Walls are not only between sides, and allow for mixture, are functional, and have thickness – they are also *dynamical.* Like those border guards, walls actively respond to whatever happens on either side so that they shield one side from the other, for the most part holding in what is on each side. A grounded copper cage serves as an electromagnetic shield by rearranging its electrical charges (namely, currents of electrons) over its area and within its thickness. If things change outside or inside, the cage's charges move around so as to cancel or modulate the effect of those changes on the other side. Changes in the internal or external temperature will require a thermodynamic wall to respond appropriately to maintain whatever conditions it is fulfilling. The wall may do its dynamical work on its own, as in the grounded electrical shield, or perhaps require our assistance, as in maintaining a constant temperature wall.

WALLS FOR A FACTORY: A PHENOMENOLOGY

Whatever everyday walls do, physicists have to make their walls do the technical work of manufacturing Nature. Physical walls do this technical work through quite detailed mechanisms or physical interactions.[13] But in abstracting and adapting everyday notions of walls, physicists are up to rather more phenomenological tasks. For example, their walls have to separate and shield.

So, whatever the kind of the wall (boundary, interface, functional, skin, dynamical) and whatever it does (delimit, be permeable, maintain conditions, bind, or respond to the world) – what it must do is *separate* one side from another. Now so far I have been describing walls as if we could see both sides. But, in fact, we or our apparatus are often on one side of a wall, and at best we can burrow into it. And usually we are on the outside. If a wall is designed to separate, practically that means it controls what we can see of the other side, the inside. Now, in actual fact, if somehow we do have a chance to get to the other side, taking a much more intimate look at it, we are often overwhelmed with the complexity of that other side. What we usually do not see, and about which we would otherwise have little if any inkling, is often more than we want to handle. Walls do the work of manufacturing Nature, and they simplify Nature so that physicists can make sense of it. Put differently, in a division of labor the factory owner need know (and may be pleased to know) only very limited features of each laborer.

Moreover, walls account for the persisting identities of objects. From the outside, an object will appear to possess enduring qualities. No matter how we look, it appears the same. Yet under much closer examination, it may well turn out to be changing inside, none of which changes are ordinarily seen by us. We can say either that the walls hold the changes in or that the walls prevent us from seeing those changes; phenomenologically, they are the same. Walls are said to *shield* many degrees of freedom, so that those degrees of freedom cannot express themselves and so they are not felt by outsiders (or othersiders). In effect, most of what goes on inside cannot show itself to us. And walls also shield against our own actions. They do not allow us to get a peek at or to get hold of most of what is inside. We cannot penetrate the wall, at least in these ways. Any time we try to influence the other side in these ways, so as to find out about it, the walls shield that other side by dynamically working against our action – think of a bulletproof vest. Of course, walls may be breached by energetic impacts or will be briefly penetrated by fluctuations (classical or quantum).

Now, in imputing a factory-design to Nature, a division of labor that produces Nature, the walls that separate and shield turn out to do so by being able to (1) filter degrees of freedom, (2) define nearbyness and own-or-other (or friend-or-foe), and (3) deal with fluctuations. And then, as we shall see, Nature will turn out to be simple, symmetric, and stable – a form of Nature a physicist could take hold of.

(Some readers may want to skip ahead and return later to the details of this section. Again, what is perhaps most interesting here is the nature of the physicists' concerns, not the exact details of how they are fulfilled.)

(1) To separate and to shield is to *filter* the degrees of freedom and so provide good handles. A wall hides or filters out many degrees of freedom. But, most crucially, it lets through or displays a few degrees of freedom which epitomize what we might call internal features of what is otherwise hidden.[14]

For example, the usual properties of a gas of atoms depend on walls that filter out or, say, average out particulate properties, nonuniformities, and fluctuations. And so they transmit what are called bulk thermodynamic variables, such as temperature or pressure. If we are running a steam engine or studying the atmosphere we do not want to know about every atom in the steam or gas. But the temperature and pressure are crucial features. (Technically, here the wall is both the experimental or mechanical setup we employ and our looking for those bulk properties.)

A balloon considered as a wall filters out transient nonuniformities in the density of the gas it encloses. When the balloon is inflated, the balloon material is fairly rigid and so is appropriately unresponsive to small transients. It does not change much. And, just as surely, the balloon shows the average pressure of the enclosed gas.[15] Now, there might well be peculiar transient arrangements of the atoms of a gas in a box (say all the atoms are in half the box, the other half being empty). Taken as a wall, what a box does is to ensure that the effects of such peculiar arrangements are drowned out by others that are both much more likely and more uniform. And so we get the conventional degrees of freedom or handles for a gas in equilibrium: pressure and volume and temperature, so that pressure times volume is proportional to temperature, the ideal gas law. A filter not only lets through a few degrees of freedom but, by the filter's actual construction (a rigid box, for example), some of the degrees of freedom that are let through are so to speak created – for we would not have gotten hold of them without that filter.

Of course, a wall as a filter is only as good as the kinds of assault we are allowed to make upon it and the sensitivity of our probes. The wall must be resilient and responsive to ordinary assaults, giving but not breaking within the usual range of insult. But if we probe a surface with a blunt yet forceful tool, or a very pointed one, we shall not only get through, but rupture the surface as well; and if we are allowed to heat up an object sufficiently its protective shield will vaporize away. Physicists design walls, or can find walls in Nature, that are in just the right balance of filtration, resilience, responsiveness, and permeability.

(2) In this factory called Nature, not only must the walls separate, shield, filter, and be resilient, they must divide the world into places that are "nearby" each other and those that are far from each other – namely, on the other side of the wall. As we shall see, the grammar of nearbyness is technically a matter of connectivity, shared properties, and correlation and symmetry, while phenomenologically it is a matter of own-and-other. These technical and phenomenological demands shape theoretical constructs in not-so-subtle ways.

We might imagine a wall that separates a uniform medium into two sides, but there is otherwise no difference between those sides. Then two points are on the same side, if we can go (by some allowed path) from one point to the other without hitting the wall. Now, if there are properties that differ sufficiently so as to distinguish the two sides – whether they be the density of a liquid vs. that of solid, or the presence of charge inside a particle's wall vs. the absence of charge outside – then rather than pathfinding, we might measure those properties to find out whether we are on the same side as another point, and perhaps which side we are on.

Whatever the properties, what happens on one side of a wall is likely to have a more profound influence on that side than whatever influence it may exert through the wall to the other side. There is greater correlation among aspects of the same side than there is between sides. So, again without being able to see the wall, we might believe we can tell whether we are on the same side as another point.

Often, delineating nearbyness requires not only separation, but the creation of an inside and an outside, effectively a closed shield; namely, no matter how we approach something we encounter the wall, and so that thing looks essentially the same no matter how we approach it.

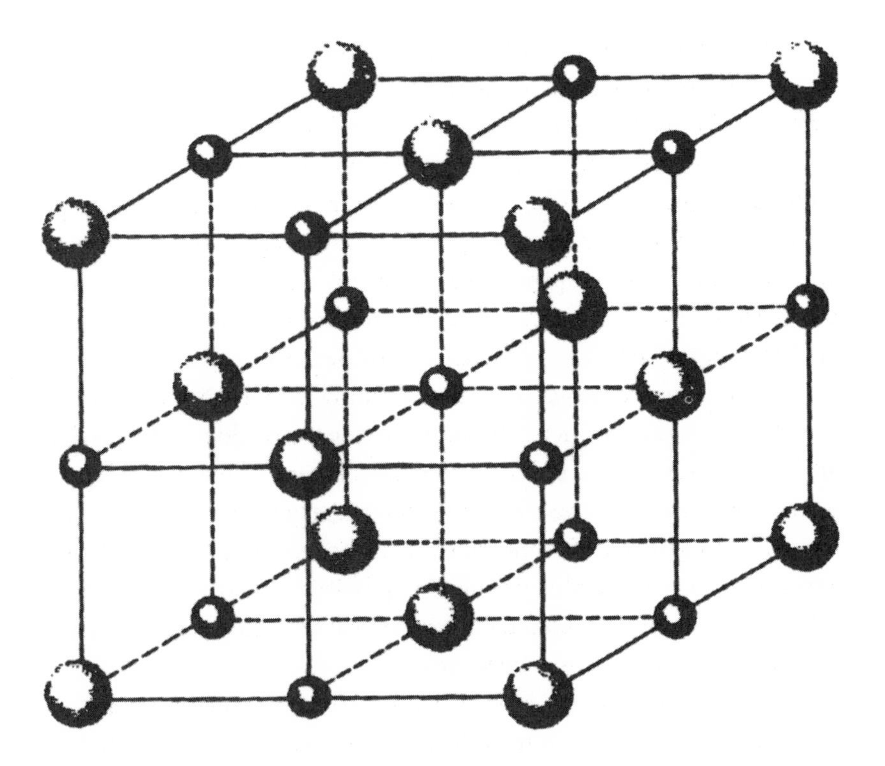

1.1. A crystal lattice

There is no back door to the insides. Conventionally, here no-matter-how means "from any angle," which is then taken to say that the object is like a ball, that it has no missing backdoor parts. But then physicists generalize no-matter-how so that it specifies a range of ways of trying to get into (or out of) something, and so a range of ways in which the thing looks the same no matter how we approach it. The geometrical image of enclosure is generalized to a functional one. So, for example, no-matter-how is taken to mean "with whatever momentum we approach something," which can be shown to imply that the object is in effect like a point – and so it is impenetrable; it could only possess properties that require no spatiality. Or, no-matter-how may mean something looks the same independent of whether the probe we use is a neutron or a proton. This says that the wall is composed of material that exhibits mostly the nuclear force, a force that does not distinguish the neutron from the proton; it is invariant to (or remains unchanged by) a neutron-proton interchange (what is called "isospin symmetry"). Physicists call these no-matter-how's the symmetries of a system, its invariances and conserved quantities: If we look in any of these equivalent or symmetrical ways, we won't get through the wall and we won't be able to change the insides, and hence there are unchanging or conserved quantities that characterize the insides. In this sense, good walls exhibit Nature's symmetries.

Consider a crystal with its highly regular arrangement of atoms: They repeat in lockstep fashion, and the crystal structure may be seen to point in a direction in space. No matter how we shine light such as X rays onto it, what we will see are reflections off the crystal's regularly spaced planes, modulated by effects due to the details of the atoms within a typical unit crystal. We cannot see the crystal's individual component atoms one by one. In effect, the regular arrangement is a wall.[16] Put differently, individual atoms are stuck in a set of interchangeable places, literally not being able to get out of the enclosure defined by that regular arrangement.[17]

For a physicist, walls simplify the world, taming its degrees of freedom – not only by conventional filtering but also by providing for symmetry or equivalence or invariance (those no-matter-how's), actually another way of thinking of filtering. And, as we shall see, these features of separation, shielding, filtering, resilience, and symmetry are built into the formal technical structure of much of theoretical physics so that they are present automatically, so to speak.

Phenomenologically, another definition of nearbyness is in terms of *own-or-other.* Walls can delineate objects so that objects are either like or unlike each other, friend or foe: "own" or "other" to each other. One side or an inside can sense ownness, so distinguishing itself from others and outsides.[18] Here, objects must have just enough distinctiveness so that each object may be distinguished from others yet not be distinguished from itself (by redundant or "degenerate" labels, for example, labels only apparently different from each other, perhaps drawn from another point of view or experimental setup). Moreover, an object may be a collection of parts, as a gas is a collection of molecules, or a family of elementary particles is a collection of individual particles. When considering a particular gas or family, one wants to be sure not to distinguish it from itself by employing walls that are too fine in their sensitivity, finding differences which are irrelevant or only apparent. To get the walls right is to get the degrees of freedom right is to get own-or-other right. And ownness is often a matter of what scale or fineness of resolution we are working at, for if the resolution is too fine own will seem other to itself. We'll see differences that are taken as artifacts of how we look.

We want to design the factory of Nature so that ownness means integrity means enclosedness means difference from other – automatically. For physicists, such walls have to have three features. They must allow for enduring charges or properties (an integrity, so to speak), for comparison, and for stability.

First, in order to be own, to have an identity, a side or an object has again to exhibit some persisting and invariant properties, such as temperature, charge, or mass, properties which are the same no matter how we look. Then the object won't be distinguishable from itself, and it is independent of us. Those properties of an object that do vary depending on our point of view ought to be accounted for in terms of the point of view. So the phenomenology and grammar of ownness might be said to place requirements on the technical physics.

Second, if own is to be distinguished from other, an object must be able to be compared with all the objects (and they with it). This might be taken to mean that the walls of an object are permeable enough for an object to sense the rest of the world, so that it could be testably other to those objects (presumably other than itself). Or, we might imagine each object potentially being moved over adjacent to other objects – being sure that that moving over does not change things – and so allow for a comparison by overlap and identity.

Third, and finally, whatever the effects of the wall's permeability (and, as we shall see, its fluctuations), in the end those effects are still small, marginally affecting a wall's thickness, even leaving a charge unchanged. Otherwise, as a consequence of permeability there would be poor separation, making those comparisons meaningless, or the wall would readily break down, or the object's properties would no longer be invariant to how we look.

In thermodynamics, for example, these three requirements are intimately connected. They define objects which have well-defined temperatures (thermodynamic systems, such as a gas in equilibrium), walls that allow for gently changing the object,[19]and equilibrium, respectively.[20]

Highly technical features of physical theories can be justified by these phenomenological requirements we might place on walls, and that is why I go into such detail about own-or-other. Practically, walls in thermodynamics and in electromagnetism are defined in just these terms. Moreover, it would seem from the grammar of what we mean by walls that there are lots of reasonable sets of tasks that walls do (own-or-other, hold in degrees of freedom, conserve charge), and so there might well be different physical models and theoretical structures to account for each of these sets. Yet, in the end, as we shall see, the grammars of the different tasks might be shown to imply each other, and the physical models might be shown to be formally equivalent in mathematical-physical terms. Still, how a wall is defined physically – which definitions and phenomenologies are emphasized – depends on what Nature will allow. The variety of in effect equivalent grammatical conceptions of walls is no intellectual curiosity, but is taken as a significant fact about Nature – for a physicist. Only certain mathematical and formal theoretical structures will do this demanding work. Own-or-other is a physical as well as a grammatical and an ontological fact.

(3) Now, in actuality, walls can filter and simplify only so much. Again, there is unavoidable leakiness, imperfection, permeability, and brittleness, and so the wall might be breached and we get inside. Put less insidiously, there are *fluctuations.* Namely, what is on one side almost always can get out to the other, at least in part, although it then might go right back in. Walls seem to have transient holes, openings that appear and disappear. So there is a recurrent sequence of intermingling and separation. The inside and the outside, or one side and the other, intermix; and degrees of freedom that are supposed to be hidden on one side may exert a subtle influence on the other side.

As in the cultural world of race and gender and pollution, in the physical world there is no way to completely avoid mixture and fluctuation. And this fact is enshrined in the second law of thermodynamics, which assures us that there will always be energy fluctuations in matter, and in the Heisenberg uncertainty principle, which says that there is a minimum nonzero energy fluctuation even in empty space.[21]

Still, the physicist tries to design walls that tame such mixture and fluctuation. For example, if the purportedly hidden degrees of freedom do exert an influence due to fluctuation and mixture with the outside (which, of course, might well disturb the own-or-other distinction), their effect might be epitomized and so domesticated by a change in the object's mass, or in the thickness of its skin (in effect, making it fuzzier, or thinner, or more stretchy), or in its actual size (in effect, perhaps making it bigger, since it extends itself into the outside). More generally, the permeability and absorptivity of a wall may well be altered; but then the integrity of the object is reasserted.[22] On the other hand, if the fluctuations were to have no effect, even though we know they must occur, physicists need to explain why; for example, why the fluctuating flows in each direction just balance exactly.

Finally, if the fluctuations are quite large, intermingling and mixture lead to a breakdown of the wall, namely, melting.[23] Different sides, once characterized by different properties, become less distinguishable. Again, phenomenologically, if fluctuations are large it is no longer so clear that we are separating the two sides or that own and other are different.

WALLS AS PROVIDENTIAL

In sum, physicists' walls are both *dynamical* and *topological.* Dynamically, walls employ mechanisms that filter yet fluctuate, that shield, coat, surface, bind, and interface – so defining the influences that can be felt through the wall and those that cannot. And, topologically, walls define separation and nearbyness, and inclusion and exclusion, delineating how well connected two points are to each other. If there is enclosure, insides have limited influence on outsides, and the insides are finite in size; and, of course, it is the insides that are enclosed. And outsides have limited influence on, and so knowledge of, insides; and as outsides, they may go on forever (to "infinity"). And, finally, walls are defined by patterns or symmetries: patterns of influence they resist, and patterns of nearbyness or identity.

Again, what walls must do must be done by actual physical interactions. Nothing else will do the work. So walls acting as shields restrict the influence of most of the many degrees of freedom of a side (say, an inside) by literally counteracting them.[24] And walls, often through elaborate cancellation effects (see chapter 4), allow for the appearance of just an epitomizing handle or two, such as mass or charge or spatial symmetry, that provides predictable and controlling access to what is inside. Put differently, walls, whether they are conceptual or experimental, are a compromise with Nature. They simplify and tame Nature, and so make Nature something onto which physicists can get a handle.

I have described walls as functional, grammatical, and physical. Functionally, walls have to do certain things. Grammatically, the things they must do define what we mean by a wall – those separated sides – and so we are allowed to refer to a wide range of mechanisms as walls, even if they do not look like conventional walls or surfaces. And, physically, walls involve actual mechanisms and processes that do the work that walls must do. So, in dividing up the world by means of walls, physicists have to say (at least implicitly) what a wall must do, what a wall is, and how it goes about doing that work. They try to invent walls that fulfill these requirements; and they believe they have discovered real walls if, again, the walls fulfill *all* these requirements. Physicists work at making sure the walls they invent are good enough to allow them to do physics – that particular conventionalized scientific activity employing a tradition of archetypal ways of doing business. Walls have to be properly impermeable, two-sided, with perhaps an inside and an outside, showing just what needs to be shown of the inside to the outside, hiding what is irrelevant. Consider this description, by Richard Feynman, of how it is possible to do thermodynamics even though its postulates are denied by everyday life:

> How then does thermodynamics work if its postulates are misleading? The trick is that we have always arranged things so that we do not do experiments on things as we find them, but only after we have thrown out precisely all those situations which would lead to undesirable orderings. If we are to make measurements on gases which are initially put into a metal can [a wall], we are careful to "wait until thermodynamic equilibrium has set in," (how often we have heard

> *that* phrase!) and we throw away all those situations in which something happens to the apparatus, that the electricity goes off because a fuse is blown, or that someone hits the can with a hammer. We never do experiments on the universe as we find it, but rather we control things to prepare rather carefully the systems on which we do the experimenting.[25]

The field of thermodynamics depends on setting up the right kind of walls (say constant temperature, or impermeability to heat), for then we can say something "physically interesting" about a system. Even if something does not seem to be wall, but functions as a wall does – as does the orderliness of a crystal – treating it as a wall or shield may allow us to find interesting phenomena, such as the gentle disturbances of that crystal structure (that wall) that account for many properties of solids. In general, if designed properly, such constrained, cleaned-up systems may well correspond to situations that are of wider practical interest. This achievement of an applicable abstraction is the real triumph of physical science.

Now we might pose one of those marvelous questions about Providence: How could Nature's walls do so many things?[26] There are, I think, two sorts of sensible answers. First, if Nature's walls did not do all these things, they would not be analogized to the more mundane brick and plaster and wood separations that make up our homes. We would not call them walls; Nature's walls would not have the right grammar. (And, of course, Nature's walls probably do not do everything everyday walls do.) Second, we might develop a physical theory of walls which connects their many features and shows how they are mutually necessary and compatible. Thermodynamics, statistical mechanics, and theories such as Maxwell's electromagnetism are in part attractive as theories because they do give such a comprehensive account.

Still, once we have figured out how the walls are to work, we retain some room in which to maneuver: namely, in what we take to be the particles and the fields – what the walls enclose and what is left over to make its presence felt everywhere. The connections between the particles' charges and the fields place further constraints on our design. And, of course, the division of labor must be done so that the factory (in effect, the theory) is able to produce Nature as we know it.

III

PARTICLES, OBJECTS, AND WORKERS

As long as the walls are doing their work, physicists do not think much about walls, nor need they do so. For the walls set up what are taken to be the natural and experimentally useful divisions of the world, and those walls in effect disappear behind the action they permit. (Now, reverting to the physicist's point of view, "we.") Rather, we are concerned with what is enclosed by those walls, the rooms or particles, and with what is excluded, the outdoors or the field: what is on one side and what is on the other.

When we are outside, what we encounter is an object – an elementary particle, an atom, or a thermodynamic system such as a gas – its insides more or less well shielded from our inquiries and knowledge. It is something like a black or grey box, perhaps fairly rigid, often something like a billiard ball or an apple. (Of course, it might be more fluid and flexible, pudding-like, something like a Claes Oldenburg soft sculpture.) Such a billiard-ball-like object is often called a particle, and we shall here use that term for all enclosed objects (even those, such as thermodynamic systems, which are usually quite extensive in size). We take it for granted that a billiard ball is localized in space, having a well-defined size and location; stable, and so it is unlikely to fall apart; nicely defined by its position and color; separable from other such balls; and capable of interacting with them (say, by bouncing off of them).[27] That is what a ball has to do and be. Whatever it is made up of, whatever is inside, however rough or resilient its surface and fillings, a rather small number of features characterize how it behaves for the most part. Its so-called inner details, if there are any, are mostly hidden from us, unless we want to cut it open or crush it or drill through it.

Much in the same vein, the individual factory workers who participate in the division of labor are known traditionally by their properties: their skills, mobility, wage demands, and so forth.[28] And we take it for granted that they may be localized in space *as* workers (perhaps ignoring their family and place ties), that they are capable of doing their work with some resilience and persistence, and that they are distinct individuals who will work in coordination with other such individuals. They are meant to be and act just like any of the physicist's Cartesian particles or objects that we encounter from the outside.

Now, the creation of modern economies required a great deal of social dislocation and reorganization – a process of alienation – so that persons would act as individuals appropriate for the factory, rather than as members of an estate or a farm or an extended family. Similarly, as the quote from Feynman indicates, a good deal of device may be needed to set up the particles that are supposed to do some of the work of manufacturing Nature. We have to learn to use walls to do the right sort of enclosing, so as to create the right sort of individual or particle or billiard ball, with the right properties and means of interaction – "right" in that they produce Nature in our factory (or theory or experiment). We need to find the right environment or external conditions so that an object can act like a particle and not fall apart, its walls vaporizing away. Again, think of factory owners trying to recruit and train workers. Once we have created a factory system of isolated but coordinated workers, or a model of Nature in terms of walls and of well-defined and separable but interacting particles, we again take such an ontology as natural and obvious, oblivious to the great deal of poignant and obsessive effort required to achieve such an organization.

WHAT PARTICLES MUST BE LIKE

If particles are to act like particles ought to act – namely, somewhat like those billiard balls and somewhat like those factory workers – physicists will say they must be, or at least implicitly demand that they be (1) localized and separated, (2) stable and objective and distinctively named, and (3) individual and additive. That these features are the crucial features of being particulate – for physicists – says how physicists come to interpret what a billiard ball and a worker are like.

(1) Particles have to be *localized,* having a position and a characteristic size or enclosing radius. And so it is possible to speak of penetrating their surfaces and getting inside of them (if they are not point-like). But localization is not only in ordinary geometric space. By artifice and convention it may be achieved in other ways. Recall, for example, the

wall provided by a crystal's regularity and orderliness. In a crystal lattice some of the right particles are in fact not the individual atoms that it is composed of, but literally the vibrations of the lattice of atoms as a whole. Such vibrations, considered as particles, are then said to be localized because they have a fairly definite frequency, momentum, and velocity (for "phonon" particles, the velocity of sound).[29] Moreover, those crystal vibrations will bounce off of other vibrations, much as billiard balls bounce off of each other. Or, consider another example: In accounting for superconductivity, the right particles are composed of two electrons acting in concert within a crystal lattice (Cooper pairs) – so we develop a formalism that allows us to treat the two electrons as a single object suitably localized in a peculiar space. The wall here is both the mechanism within the crystal lattice that links up the two electrons so that they act as a single unit, and the mathematical formalism that enshrines that pairing (and which, as we shall say in chapter 4, correctly defines the vacuum, its symmetries, and its excitations). As we shall see, physicists seem to believe there must be a way of looking at almost anything so that it is localized – somewhere.

Now each localized object must then be *separated* from other such objects. There must be the possibility of space between them. The trick is to find a "space" (again, a mathematical-physical space, which may well not be a matter of ordinary extension) in which particles we take as distinct are in fact separated as well as localized. So when physicists introduce new quantum numbers (such as charm or strangeness), they are identifying an enlarged space in which differing particles are now differently labeled and so they are separated and distinctive. As we shall see later on, finding a good mode of separation (such as an enlarged space) has proved to be a pressing problem for certain particles that are too much alike for comfort, such as the muon and the electron. Moreover, as we show in chapters 2 and 3, once physicists find such a space good enough for separation, they then have an understanding of the orderliness and symmetries of Nature.

Again, note the kind of description I have been giving: Physicists believe particles should have certain properties. If something looks like it might be taken as a particle, one finds a way of so taking it (namely, a good space and a good experimental setup), a world in which that thing acts particulately – "just like" a billiard ball acts in its world of the billiard table.

(2) Particles are not only localized and separated; they are also stable and objective. Here we parallel the description of walls. To be *stable* is to be able to withstand a range of insult and still remain roughly the same.[30] Again, the trick here is to restrict the sources of insult, namely, to set up both an external environment that is tame enough and the right sort of wall so that the putative particle actually is sufficiently stable – if that is possible. At the same time, physicists want to justify that environment, both by its being a sensible restriction of the range of insult and its being a reasonable environment in which to work. So, for example, we might restrict the range of temperatures to which the particle is subjected (so it won't melt) and at the same time note that that range is a practically useful one or one that naturally occurs (as in room temperature or, notionally, as in the Big Bang). Similarly, in the factory, one sets workloads and working conditions so that individual workers act and believe they are individuals who can actually do the work – namely, "scientific management." Again, what we as physicists or manufacturers are up to is creating and justifying conditions in which the phenomena we are trying to manufacture can and do naturally appear.

A particle is *objective* if no matter how we look at it, it is again the same, having an invariant set of properties.[31] Of course, the problem is to invent and justify a range of no-matter-hows. Some ways of looking – crushing or exploding – will almost surely change what we see, so we need to say which ways are out of bounds, or we need to give an account of how what we see is affected by our mode of looking. So if the back of an at first seemingly round object looks different than does the front – consider a half tomato – we might want to say that the angle of approach affects what we see, and *this* is just how. Differences in what we encounter are taken account of, and so we might say that we are seeing the same object.[32] (As we shall see, one of the functions of fields is to give such an accounting, an account of how the path to a particle determines what we see.)

If, for example, we probe a nucleus with a ballistically shot electron, then the electron, especially if it is quite energetic, will surely penetrate the nuclear surface, and so what we see will be different depending on the energy of that electron. In fact, we may so obtain a detailed look at the internal structure of a nucleus. For, if we can give an account of how that difference depends both on how hard the electron hits the nucleus and the size and shape of the nucleus itself (assuming that the electron is point-like), we might then say that we are seeing different aspects of or different depths into the same object. We take it that if a particle can be made objective in this sense – successfully separating it from how it is seen, just as a billiard ball is objective – it has the possibility of being real.[33] In this manner we have observed the distribution of charge in a nucleus and the quarks inside a proton. Now, that successful separation is an act of theoretical abstraction. But, the presumably real particle that is its consequence may then actually be probed by many other means, and still we see the same particle. Hence, for a physicist it actually is real.

Names. Nature is sufficiently peculiar so that we may need to do further work if we are to achieve objectivity. A particle may well need a fairly complex name if we are to account for its appearances and capacities under the various modes of looking at it or taking hold of it. It has more than one or two relevant features or properties; that is, it displays a goodly number of degrees of freedom or handles. Only with such a complex name will it be well distinguished or well separated from its cousins. Otherwise, it would be the same, at least in name, as they are – yet the cousins will act differently in some circumstances.[34] And if such differing cousins are not well distinguished, there won't be objectivity. So, elementary particles not only are known by their masses and charges, but by their angular momenta, isotopic spin, "strangeness," etc. – properties needed to describe what they do.

One of the fantasies of theoretical particle physics is to need only a one-term name, say the mass, which we are able to deduce or calculate from all the other properties of a particle, or the other way around, going from mass to properties. And, as Feynman suggests, we might try to sufficiently restrict the environment, the range of insults and probes the particle will feel, and the roles the particle takes in making up Nature, to keep its name simple. But often this strategy proves to be overly restrictive on environment and role. In actual practice, we are stuck with rather more provisional and practical theories. And so we are stuck with making sure that each particle has just enough labels so that we can account for the particle's various phenomenal manifestations in the laboratory or in our messy world. Then we may assure ourselves that there is really just one object – a particle – in front of us. And, of course, these labels localize the particle as well.

Now, what turns out in the end to be the same kind of particle may in different environments or practices be called light or a gamma ray or a photon particle; another might be called doubly ionized Helium, or a Helium nucleus, or an alpha ray, or an alpha particle. And in each actual case we act differently toward it. (This is not a matter of degeneracy, but rather of the different practices in different subcultures of physics.) A particle of electromagnetic radiation when grouped with the radioactivities is called a gamma ray, going along with alpha and beta rays. But when it is grouped with the "vector bosons," particles with angular momentum or spin of one (going along with the W^+, the W^-, and the Z^0), it is called a photon and is symbolized by a lowercase Greek gamma, γ.

The natural associations of each naming scheme are different: whether it be radioactivity and irradiation; atomic interactions; individual particle detection; or taxonomy and systematics.

In sum, a name, as a set of properties that uniquely define an object, depends on the work the object does, the kinds of insult it is subject to, and the kinds of particles it is "just like" or more or less indistinguishable from – under certain conditions.[35] And since the empirical, experimental, and theoretical arenas of physical research and experience are often parochial, there may be several distinct but more or less equivalent names for a particle. Depending on which name we use, we shall think differently about it and come to somewhat different associations and implications. Even if eventually we may show the formal equivalence of different names, in practice those differences matter enormously.[36]

Names are so to speak necessary, being embedded in practices and theories. Names have modes of action (or a grammar) associated with them. Even proper names have motivations: either systematic (for example, alpha, beta, and gamma rays, here an alphabetic lexicographical system); or anecdotal (quarks and strange particles); or both (Californium and Neptunium, lambda's (Λ) and their V-shaped decay tracks).

As we shall see in chapter 3, to be "just like under certain conditions" is often a statement that we have a theoretical structure that puts particles into families. And from the relationships within a family, and relationships between families ("exogamy"), we can understand what the particles do. Moreover, those names imply regimes of stability for the particle – stability to some insults, vulnerability to others (here, the possibility of interaction or marriage).

To recapitulate: Names are meant to have consequences. With the right kind of name we then know how to maintain a particle's identity or separation, its stability or its resilience to insult, and its objectivity or its invariance to the way one looks at it. In actual practice, objectivity depends on the particulars of the context and purpose in how we look, and so there are different names that designate the same particle yet are very different in the practices they imply. The seeming arbitrariness of names is disciplined by the demand that particles be objective; yet the totemic objectivity of particles is humbled by the parochialism of names.[37]

(3) A last requirement we often place on particles, one we shall return to in some detail in chapter 2, is that particles (like those billiard balls) are *individuals*. Each particle may be considered on its own. Even though it may interact with other particles, those potential interactions, as such, do not for the most part violate the particle's stability or integrity. Some of its properties are pervasive and indivisible, say a particle being a muon, or the temperature of a gas. Other of its properties allows it to weigh in with its value, so to speak. For example, the energy of a composite system of several particles might be just about the sum of their several energies. This suggests that we might even more generally add up those particles or their properties, saying that particles are in effect demographic. What is impressive to physicists is how wide is the range of circumstances when integrity and additivity may be made to apply, so that even in complex interactive environments there will emerge particles that are in these senses stable and additive.[38] And this recurrent emergence of individuated particles confirms for physicists their faith in their notions of individuals.

More technically, there is a range of circumstances such that when we put similar particles together we expect there will be some properties of the collection – called extensive properties – which will be the arithmetic sum of the component particles' properties: such as electrical charge, mass or weight, energy, and entropy.[39] Extensive properties reflect our intuition that the billiard ball has a weight and a volume, and, again, that a collection of billiard balls has just their total weight and volume (if we can ignore the empty spaces between the balls). Reflecting our intuition of the billiard ball's color, there will be other properties – intensive properties – which, if they are the same for the individual components, remain the same for the collection: such as temperature or chemical composition.[40]

Not all seemingly extensive properties actually turn out to be additive; and so they are said to be not well defined for a system. But then physicists expend great effort to find improved versions of these properties. The total heat energy (not the total energy, as such) needed in order to change a gas from one physical state to another – a change in temperature or pressure or volume – will depend on how we supply the heat, on the path we follow. When we sum up the heat required along different paths, all of which have the same end points, we'll get different values; the effect of heat contributions depends on the conditions under which they are added in. In this sense heat content is not a well-defined extensive property; it does not cumulate nicely (unlike in adding up numbers, where $1 + 2 + 3 = 6 = 3 + 1 + 2$). What is nicely cumulative is a measure of heat in terms of temperature: namely, entropy, the heat contributed per unit temperature. Entropy is a good extensive property.[41] So thermodynamics is taken by physicists as the story of temperature and entropy, not of heat. For heat is not a right degree of freedom in this realm.

Not only do we modify the definition of properties in order to assure ourselves that presumably extensive properties are nicely cumulative, we may impute the existence of new carriers of these properties so that an extensive property remains additive and so extensive. So, in order to account for and so conserve energy (among other properties) in beta radioactivity, Wolfgang Pauli (1930) proposed the existence of a presumably difficult-to-see electrically neutral elementary particle, a neutrino, emitted along with the electron in beta radioactivity.

In contrast to the gentleness of interaction between individuals that allows for the additivity of their extensive properties, there is the pervasive coupling of an individual to itself and then to its environment that then allows for size-independent intensive properties. Internal fluctuations or variations are in effect vanishingly small. For example, when physicists attribute a single temperature to both the massive elementary particles and the electromagnetic waves or radiation in the universe at early times in its development, they are making a remarkable claim about the wholeness (actually, the equilibrium) of the universe at that time. More generally, we might say an object is stable to insult, since it cannot be gotten hold of except as a whole, at least by this handle. Hence, it really is a single indivisible individual.

Here, as elsewhere, the physics and the phenomenology go together, and they must be made to do so. If an object may be characterized by extensive and intensive properties, such as weight and color respectively, and so it is additive and stable, then it might well be a billiard-ball-like individual particle (even if the particle is an enclosed gas, considered as a whole). Physicists then justify why there might be both extensive and intensive properties: namely, how the detailed physical mechanisms allow for these phenomenal facts. For example, when interparticle or intersystemic interactions are mostly absent or weak, then charges or entropies of various sorts add. When intrasystemic interactions are stronger yet fluctuations are small, then equilibrium is possible and a particle might be characterized by pervasive intensive properties such as temperature.[42]

INTUITIONS OF WALLS AND PARTICLES

Particles are designed to be localized and separated, stable and objective and named, and individual and added and propertied – if they are to act like those billiard balls. Now if walls define an inside and an outside, and so they enclose and localize and hide, and the walls are resilient to breakdown, then they can make for a nicely handleable particle. What walls "must be" produces what particles "must be," to speak both literally and conceptually. Ours is a story of grammar, metaphysics, and physics – a story of meaning, of natural units or individuals, and of mechanism.

Now we might wonder if we are misled by our everyday conception of billiard balls or of walls as we try to make Nature fit our naive intuitions. We might well be. But what is impressive is how we modify our naive intuitions, teaching ourselves to notice the right features of these everyday objects so that Nature is modeled by them. Analogy is destiny, but it is how we analogize that gives us our destiny back to ourselves.

Similarly, one of the major political problems of modern economies is to convince people that the kind of individual needed for the economy's functioning is an intuitively reasonable kind of person. I suspect that it is the success of these abstractions, that the physicist's and the entrepreneur's factories do produce a form of Nature and lovely material objects, that makes us more willing to go along with the required revisions of our intuitions. But I also suspect that the abstractions are not so far removed from experience, and the arguments for such abstractions are not so outlandish, that we may see how we might go along with them without stretching things too far.

IV

Given walls and particles, we might have sufficient workers and capital for manufacturing Nature, a sufficient division of labor to make the factory work and even work efficiently.[43] For example, one can describe the electrical or gravitational interactions of particles in terms of how individual particles affect each other at a distance, and if those particles are not moving too fast (compared to the speed of light) we recover those famous inverse square laws (Force $\approx Q_1Q_2/r^2$, where Q refers to electric charge, and r is the distance between the charges).[44] Still, for many practical situations it is awkward to try to get away with so meager a set of work roles. Rather than achieving versatility through greater complexity, perhaps the factory needs greater articulation and specialization, a greater and more supple division of labor, even if it means that we have more machinery and kinds of labor than we might absolutely need. With the flexibility provided by a larger variety of elements, it is easier to handle a wider variety of situations, and it is possible to make that handling more natural, less jerry-built or awkward – which is always the case when one tries to do everything with too few tools.[45] It is perhaps better to be less parsimonious than to feel that one is shoehorning Nature into a rather rigid and intricate production line.

To make for a more powerful and effective division of labor, physicists often need the notion of a *field,* where they have in mind something like a flowing fluid or a magnetic field. Now, our initial intuition was perhaps that all there is are shielded objects with space between them, and hence all there is are interactions of those objects. A handle onto an object, such as a charge, is in effect just a summary degree of freedom. But, physicists then say that that handle is also, reciprocally, a "source" of a field that provisions that empty space with properties or degrees of freedom, such as the electric field at each point. And such a field then affects other shielded objects through their handles or charges. In this manner, walls, particles, and fields, filling up all of space, are to be made to work together, in a comparatively small variety of simple patterns of organization, so as to produce Nature with great suppleness.[46] It turns out, and physicists want to give an account of why, that that provisioning of space with its own degrees of freedom makes for a much more manageable factory.

What is a field? The conventional picture of fluid flow recalls an ocean or river. At every point in space the water has a velocity, and those local features link up into larger-scale lines of flow. The values of the field (here, the velocities) at nearby points are in general not arbitrarily different from each other, for otherwise there would be no flow, no continuity. Also, we know that the fluid is conserved: what flows into a volume must eventually flow out (unless there is a source or a sink at that point).

Fields are antithetical to particles. Being spread out in space, they are not localized. But they do have a well-defined value at each point, such as the velocity of flow.[47] Hence, fields have an infinitude of degrees of freedom since those values might be arbitrarily different at each point. But if there is continuity and conservation, that arbitrariness is severely restricted. As we shall see, insofar as fields are explicitly designed so that they work along with walls and particles, that design even further restricts and so tames their degrees of freedom and their arbitrariness.

WHAT FIELDS MUST BE LIKE

Walls are like shields and skins, and particles are like billiard balls, and fields are like fluids in motion. But in what senses are fields like these fluids? As I have indicated, the features that physicists abstract from such a fluid flow that then become definitive of a field are: smoothness, connection to particles which are its sources, and rules of accounting for sources and flows. Where space was once Kantian, the possibility of separation, now it becomes the fabric which connects all into a whole.

A good field is not only continuous, the value at each point depending on nearby values of the field itself, but the field tends to be reasonably *smooth*. The fluid's velocity changes slowly in space and time, except at well-defined places.[48] If we knew both the rules of spatial dependence or interconnection and the value of the field at the boundaries at a particular time, we might even be able to use those rules to figure out the field values everywhere, working our way in, away from the boundary, and from that particular time forward, so to speak.

A field is also *connected to particles*, particles which act as its sources, much as a fluid flowing in a river is connected to its springs and mountain sources. The properties of a particle – for example, charge or mass as ways we take hold of it electromagnetically or gravitationally – are to be thought of as representing the "leftovers," what is not fully shielded or held in by the particle's wall, what is not held back by dams and mountains and so flows out. (Note that a fully held in or shielded particle would have no handles and would not be part of Nature for us. Leakage and handles make for the world as we know it.) The field is just what leaks out or spews forth, and those particle properties represent the source of a flow, the sources of the electrical and gravitational fields.

Particles at different locations need not be described as interacting directly with each other, action at a distance, to use the physicist's term of art. Rather, at its location a particle interacts with the field of another particle (namely, what the first particle "feels" at its own location). More precisely, it interacts through its properties or degrees of freedom, just what has leaked out of it, with what has leaked out of and spread from that other, presumably distant particle. (In quantum field theory the leakages or fields may be represented by a flow of "virtual" particles, called virtual because they are not free and so "real" but are bound to their sources. For example, virtual photons emitted by charged particles account for Coulomb forces. Here, interaction is between particles, real and virtual, at coincident points. Forces between distant particles are conveyed or "propagated" by mediating particles.)

A field must *account for itself*, just as we account for our resources by financial accounting. In cartoon, accounting plays upon the difference between credits and debits, allowing for future payments and for bills that will eventually come due. It also may assume that the accounts of independent subdivisions of an organization balance separately, and that the account of the whole organization is merely the consolidation of the accounts of the subdivisions. Of course, the real problem is to design an accounting system that fulfils these ideals, if that is possible at all. Correspondingly, for the Factory of Nature, once we have delineated a relationship between source and field, then at each point the inflow of the field must balance the outflow unless there are sources or sinks of flowing material. And so the field is conserved. Here, the problem is to keep track both of the particulate sources of those fields (such as mass and charge) and of the inflows and outflows. Good sources are individuals, so that the field of two sources is just the sum of their two fields (what is called linear superposition). What leaks out of each particle adds up, point by point throughout space.[49] If fields are going to work with particles, physicists – accountant-physicists that they are – demand that the fields work in these balanced-book conserving and additive ways. They know how to handle such laborers. And so they must find a system or organization of Nature that fulfils these principles.

A good field not only flows smoothly and is connected to or is "wired up" to its sources so that it accounts for itself, automatically conserving the flow so nothing appears out of nowhere.[50] The field also *objectifies* its particle sources, taking into account their peculiarities. Say, for example, that the particle seems not quite so objective, appearing differently depending on the path through which it is approached. Path and history matter. Rather than an infinity of particular names, accounting for the different modes of approach, we might try to attribute this path dependence to a field surrounding the particle.[51] For example, the feelings we have about a meal at a restaurant may well depend on our experience of the traffic on the way there, that traffic (a field of intensity and velocity) being due to all the people who are going to restaurants. The restaurant is said to always make the same meal, even though our experience of it will vary. Or, consider a temperature field which accounts for the flow of heat (it flows to points of lower temperature) and also objectifies particle-sources in thermodynamic systems, allowing for the definition of an entropy.

Rather than saying that a particle's property (or the meal's quality) is arbitrary or dependent on how you look, one would say that the observed property depends on the values of the field along the path of observation. The particle stays simple and localized; the field also stays simple. The dirty work of accounting for differing observations is literally a product of our particular path through the field.

In sum, for physicists, fields represent the unshielded or leftover degrees of freedom, which then become a distribution or a provisioning of properties and degrees of freedom to space. Fields also provide for local interactions by, in effect, transporting the effect of one particle over to another particle with which it is to interact. (Note that fields may also be the agents of walls, here performing the tasks of walls that provide for own-or-other, such as integrity and comparison.) And, as we shall see later, fields can be given natural mathematical forms that automatically and by the way take care of conservation and flow rules, nearbyness requirements, and path dependences and additivity.

Such an intuitive conception of fields, of particles, and of their intimate and mutually constrained relationship is quite remarkable, not at all initially obvious when we examine a flowing fluid. For example, are flows just due to sources, and sources merely the consequence of flows? Could there actually be a division of labor that makes each role or task feed nicely into the others so that the factory works smoothly?[52] Here is the physicist John Wheeler's description of electromagnetism and gravitation:

> The one central point is a law of conservation (conservation of charge; conservation of momentum-energy).
>
> The other central point is "automatic fulfillment" of this conservation law.
>
> "Automatic conservation" requires that source not be an agent free to vary arbitrarily from place to place and instant to instant.
>
> Source needs a tie to something that, while having degrees of freedom of its own, will cut down the otherwise arbitrary degrees of freedom of the source sufficiently to guarantee that the source automatically fulfills the conservation law. Give the name "field" to this something.
>
> *Define this field and "wire it up" to the source in such a way that the conservation of the source shall be an automatic consequence of the "zero boundary of a boundary"* [technically, a mathematical schema that is a form of the Fundamental Theorem of the Calculus or Stokes law, a rule of differential geometry]. . . .

> One view: Source is primary. Field may have other duties, but its prime duty is to serve as "slave" of source. Conservation of source comes first; field has to adjust itself accordingly.
>
> Alternative view: Field is primary. Field takes the responsibility of seeing to it that the source obeys the conservation law. Source would not know what to do in the absence of field, and would not even exist. Source is "built" from field. Conservation of source is consequence of this construction.[53]

Now we might not be able to design a factory that would fulfill all these aesthetic and ideological demands and still produce Nature. What is worth noting is that it seems we can – if we think of a fluid in the right way, if we make the right abstractions. And, of course, we have learned to think of a fluid in just these ways as a consequence of our trying to figure out how Nature works in terms of the division of labor in such a factory.[54] If we could not get away with these abstractions, if they did not prove fruitful, then we would have abandoned those demands for ones we could get away with.

Although I have presented fields as the last component of a triplet of laboring roles, there are other factories in which fields are the leading or independent or only laborers, with phenomenologically observed particles perhaps being merely peculiar topological arrangements of field (just as in the particle realm fields are thought to represent the effect of particles). But for the most part any such monotheism has proved much less effective than polytheism and a consequent division of labor. Now, historical study does show that the institutions and divisions of labor we now take as necessary were once considered arbitrary, with other perhaps even superior possibilities now having been forgotten.[55] Yet alternatives are not easily installed or even recalled now, because our ontological commitments express themselves pervasively. We are in effect locked in.

So far I have described the laborers and the schematics of their interactions. But the system of production awaits further elaboration in later chapters. The division of labor is intimately connected with the story we tell of manufactures, how productive things are made up of other things (chapter 2), the story of society, namely that of freedom and necessity (3), and the story of entrepreneurship, how something arises out of nothing (4). More practically and instrumentally, in chapter 5 I describe how the particular division of labor, and the modes of analysis, classification, and creation we commit ourselves to, have consequences for how we get a handle onto the world, feel it out, observe it, and change it. And in the sixth chapter, we discuss how all of this is embedded in a language, mathematics, that would seem to have its own autonomy and so tells us about the world.

What is provided here is an anthropology of scientific accountings, a description of a culture. It is a philosophic anthropology. What are the ontological commitments that are assumed when we take Nature as something subject to our analysis? Just how is the factory – Mandeville's infamous bees in their hive – the logical or transcendental condition for Nature? To repeat, a schematic division of labor does not yet tell in detail how the factory works or what it produces or just how it is efficient. Moreover, for physicists the world is unavoidably empirical. If analogy is destiny, and metaphors are authoritative, just how an analogy is destiny and a metaphor is authoritative are matters that we must figure out for ourselves and check out and discover in the world.

TWO

Taking Apart and Putting Together: The Clockworks, The Calculus, and the Computer

The Right Degrees of Freedom; The Clockworks and The Calculus. Parts Are Strategies; Independence and Randomness; Dependence, Spreadsheets, and Differential Equations; Additivity and The Calculus; Disjoint Functionality and Interpretability: Bureaucracy, Flow Processing Plants, and Object-Oriented Programming; Sequence and Procedure. Parts Are Commitments.

THE ARGUMENT IS: PARTS ARE DEGREES OF FREEDOM, THEY ARE strategies, and they are commitments. In the first chapter the operative model was the division of labor in manufacture and in political economy. Now we ask just what kinds of individuals are suitable for a factory or for an economy of Nature. The operative model is a mechanism such as a clockworks, or anything composed of systematically associated or interdigitated parts, parts being the generic term that implies that individuals make up or compose something. (Note that the factory is now such a mechanism.) We shall be concerned with the process of building or making such a mechanism and, in chapter 5, the ways of an artisanal craftsmanship employed in discovering the mechanism's workings. In this mode of thinking, if we can find good analytic units – parts – then we hope that their mode of composition will be manifest, or almost so. And if we can put the world together out of those parts, those parts should be good handles onto the world; namely, they are or they possess good degrees of freedom. (Of course, the mechanism may possess good degrees of freedom not readily reducible to the parts' degrees of freedom.)

The kind of parts we choose leads to a specific strategy for analysis and composition: The probability calculus provides a strategy if the parts are statistically independent of each other, as are coin tosses; spreadsheets or differential equations, if the parts are gently and explicitly dependent upon each other; the integral and differential calculus, if the parts are to be added up; bureaucracy and systems analysis, if the parts are connected to each other through inputs and outputs; and algorithms and procedures, if the parts' modularity and sequence are crucial. While I have surely not exhausted the kinds of parts and their attendant strategies, these parts and their everyday models cover a very wide range of situations – if the way the world must be understood is through partitioning. For the physicist, the important question is what kind of parts will work to put the world together?[1]

I

> Look round the world: Contemplate the whole and every part of it: you will find it to be nothing but one great machine, subdivided into an infinite number of lesser machines, which again admit of subdivisions to a degree beyond what human senses and faculties can trace and explain. All these various machines, and even their most minute parts, are adjusted to each other with an accuracy which ravishes into admiration all men who have ever contemplated them.
>
> – DAVID HUME (1779), *Dialogues Concerning Natural Religion*[2]

> [On combinatory play, as in jigsaw puzzles . . .] Man seeks to form for himself, in whatever manner is suitable for him, a simplified and lucid image of the world and so to overcome the world of experience by striving to replace it to some extent by this image. . . . Taken from a psychological viewpoint, this combinatory play seems to be the essential feature in productive thought – before there is any connection with logical construction in words or other kinds of signs which can be communicated to others. [The jigsaw puzzle is a very good tool for learning, long before children can put the idea into words, that by arranging seemingly disconnected pieces in the correct order, one can arrive at an image of wholeness which is much more than its parts. – Bruno Bettelheim]
>
> – ALBERT EINSTEIN, quoted by B. Bettelheim[3]

THE RIGHT DEGREES OF FREEDOM

The physicist Steven Weinberg has described a Third Law of Progress in Theoretical Physics: "You may use any degrees of freedom you like to describe a physical system, but if you use the wrong ones, you'll be sorry."[4] The degrees of freedom of a system are ways of getting a handle onto it. If you choose a right one you'll have a convenient and specific handle, one in accord with the orderliness of the world; if you choose a wrong one, you'll drag things along every which way. And with the right degrees of freedom, the equations you write will be perspicuous, reminding you of valuable analogies; with the wrong ones, the equations will make no manifest sense and seem impossible to solve.

It is not at all obvious which are the right kinds of degrees of freedom for getting hold of a system. Nature goes on as she will; but insofar as we want to manipulate, control, or understand her (and the gendered pronoun here is traditional and significant), we had better use the right degrees of freedom – or we'll be sorry.[5]

For a billiard-ball-like particle, the right degrees of freedom are often its position and momentum and angular momentum; and for an elementary particle, such as an electron or a kaon (such as the *K*-zero, K^o), they also include its strangeness, charm, etc., as well, degrees of freedom that reflect its composition (in terms of quarks and gluons, as parts). For a superconductor, the right degrees of freedom are those of pairs of electrons within a crystal lattice – these are its parts. And for a system of planets going around a sun, the right degrees of freedom of a planet might be its angular momentum and its period of rotation (so-called action-angle variables).[6] Recall, these degrees of freedom are right because they allow us both to understand an object simply and to have a good handle onto the world.

Journeymen physicists can do rather well at choosing the right degrees of freedom for getting hold of the world. They do so by imitating previous strategies; their practice is conventional and traditional. From their training, they acquire a set of notions – perhaps more generally embedded in the culture – about how one should go about looking for the right degrees of freedom. These notions are the of-courses and the obvious steps, and, conversely, they include the natural errors that journeymen cannot always avoid. Again, as we have seen in chapter 1, the world "must" go this way, analogy is destiny, and everyday notions become applicable through a reinterpretation of their obvious meanings. If this is a culture of science, it is a society's more general culture that science finds itself sharing in – there is no other. These cultural notions are also employed in analyzing the economy, the polity, and social relations. (Of course, it is of interest to ask which aspects of a society's culture science does not share.)

One of the most powerful and pervasive cultural commitments is to believe that what we are trying to get hold of is made up of smaller things – its parts. Can we find a way of taking something apart into pieces so that it is seen to be made up of those putative components, that the working of the whole thing is simply a matter of the workings of the pieces together? We might hope that some of those pieces are good handles onto that thing, so that parts – actually, their properties – are good degrees of freedom. But, often the right degrees of freedom may not be attributed to any piece but rather to a group of the pieces in interaction with each other, which group itself might be taken as a larger piece. Along the way, we have to convince ourselves that we can actually put the parts together, that there is a construction that takes the analysis of a whole into parts and, so to speak, inverts it. Such is the most general story of *analysis and composition.* Here, the concrete models we have in mind are a clockworks mechanism or building blocks. Alternatively, as we shall see in the next chapter, we may specify a set of rules by which the parts interact and then demand that the parts enact every possible mode of interaction; and it is that totality of ways of interacting that makes for the story of being "made up of." Perhaps the best-known examples of this latter mechanism are the market economy and the evolutionary system.

What is fascinating is how people go about insisting on the world's being made up of parts – the strategies, the fudges, and the focusing of their concern on malleable cases.[7] There has been much written on the ideology of the commitment to such a strategy of mechanism and reductionism, and on the philosophical relationship of wholes to parts. But it is the devices and inventions, even more than the desires, that are so wonderfully apparent when we watch people go about doing their work. Craftspersons know they have to find the right degrees of freedom, they know how the analysis "must" go, and then they go to work. It is that work, and its commonalities among many areas of endeavor, that I want to describe.

THE CLOCKWORKS AND THE CALCULUS

For physicists, a clockworks mechanism and the integral and differential calculus define two very different modes of taking apart and putting together: one is a story of components that fit together, and the other is a story of the marginal addition of infinitesimals. In each case, traditional notions of mechanical philosophy are concretely embodied through the parts and how they fit together. As we shall see, physicists in finding the right degrees of freedom find parts that are taken to be componential or marginal.[8]

Components are building blocks or the wheels and gears of a clockworks. They are the usually discrete and often heterogeneous pieces found when we take something apart. We then put them together, often in a prescribed order, to make up something that then works through the interaction of those parts. So, for example, an arch is built literally of blocks, most of which are roughly the same shape but one of which is a unique keystone. An automobile is put together at the factory, and disassembled and put back together at the service station – hopefully following the steps in the right order. A complex molecule is put together out of its component atoms through a series of chemical reactions. An economy is put together out of a large number of so-called economic men, and letting the haggling among them begin.[9] At least at first, the degrees of freedom are denominated by the components or building blocks – the parts: the gears and valves, the atoms, the buyers and sellers. But each part's degrees of freedom are then rather severely limited by the interaction of the component parts – so that a clock does many fewer (but perhaps more interesting) things than do all of its components separately. And so it may turn out that the right degrees of freedom of the clock, say its rate of ticking, are not those of any part, but rather of the mechanism as a whole.

But now consider a very different kind of component, say the x and y components of a vector. Each of these components represents a degree of freedom, a way of changing that vector and presumably the object to which it is associated. Technically, those components represent invariance or symmetry, say through horizontal and vertical translation (*x* and *y* directions). And if we choose the right kind of components, the right symmetries for a problem, the parts are nicely separable. Those components are candidates for good degrees of freedom, since we might handle each of them one at a time.[10]

When analysis is conceived of as being a matter of *marginal* additions, one takes it as a given that there is a beginning point or "ground state" (whose composition may or may not be well understood in terms of analysis), and then one accounts for the world in terms of small or marginal additions to this initial state. Something is said to grow or fill up or heat, or develop or age or cumulate or accrete, or perhaps go through stages: So an integral is a cumulating sum of differentials. Or, organismic biology is in part a story of an individual's marginal development; and geology and stellar structure and cosmology are often stories of marginal development. Marginalist analysis of market economies is all about the effects of adding on a little bit more, "on the margin." And the building up of the periodic table, marginally adding on one more electron (and the nucleus's proton and neutrons that go along with it), is another such example – albeit the electronic structure becomes hierarchized into shells.

Here the degrees of freedom or handles onto the system are the directions of marginal change: size or temperature, price or quantity, or the number of valence electrons. Go from sodium (atomic number $Z = 11$) to magnesium ($Z = 12$) and see how the chemistry changes. Again, the initial point or ground state might have its own internal degrees of freedom, its own parts, but they are hidden behind walls and so they are of no concern and have virtually no effect.[11] Only the marginal changes and their degrees of freedom need concern us. (In chapter 4 we shall discuss in much greater detail the systematic hiding of such internal degrees of freedom, what is called a vacuum, and the analogous role of marginal additions as "Something," in contrast to the ground state or vacuum.)

When things are made up by means of marginal additions, each of the marginally different stages is usually whole in itself, each intermediate stage of transformation being necessary, being along the way – as in an organism at each stage of development. Composition is a matter of accretion and marginal change and development (and, in biology, duplication). The organism grows in *this* way and must go through *these* stages, and it is an authentic whole organism as it is growing day by day.[12]

Of course, what is most curious is that marginal additions do not always have marginal effects. Sometimes, slowly but surely, there is transformation, so that in retrospect there would seem to be two very different stable stages with a transition between them. Sometimes, there are rapid and discontinuous changes, straws that break camels' backs: the freezing of a liquid when the temperature is lowered just a bit, cell divisions and differentiation that lead to new organs, and new chemistries. As a consequence, there can be dramatic changes in the kind and quantity of the degrees of freedom, the handles we need to take hold of a now transformed beast, even though the addition was marginal.[13] When a liquid freezes, new parts appear; namely, there is now a crystalline order and the marginal parts include vibrations of that crystal lattice.

II

PARTS ARE STRATEGIES

Physicists employ a small number of strategies for insisting that the world is made up of parts. A strategy prescribes the kinds of parts and how to compose them, and more specifically, the parts' features we might emphasize and those we ought to ignore. A strategy tells what might appear to be alienated individuals how to be parts. Such a strategy is seen as practical and concrete, not abstract and generic – even though it might subsequently be generalized to other situations. So, for example, we build up an atom out of particular shells of electrons; and then by analogy, we build up a nucleus out of shells of neutrons and protons. As we shall survey below, notions of the clockworks and the calculus become embodied in particular models of analysis and composition: a random walk, a differential equation, the integral calculus, a bureaucracy, a computer program. Now, generically, a random walk's parts are independent of each other, while a differential equation's parts are intimately dependent upon each other. In the calculus the parts add up nicely, as do integers, while in bureaucracy and computation the effect of putting together the parts depends essentially on hierarchy and sequence and order. But, for the physicist, these particular models refer to even more specific paradigmatic concrete examples (or "exemplars," as the philosopher T. S. Kuhn might say), such as coin tossing, the heat equation, the area under a curve, a top-down military or bureaucratic organization, or an algorithm for sorting a file.

Parts are attached to strategies, and parts and strategies are attached to specific paradigmatic examples. And the examples demonstrate which kind of analysis is good enough, sufficiently explanatory, for this particular strategy. The parts and strategies and examples are just what "everyone knows" in the subculture. Now, technically, we might say that physicists deal with variances (independence), averages (dependence and flows), collectivities (addition), and hierarchical divisions and processes (bureaucracy and procedure). As we go along, we shall try to understand the meaning of such a list. But what will be as interesting is just how we insist – in particular detail – that the world must go in one of these ways. Let us see how.

INDEPENDENCE AND RANDOMNESS

At one extreme, there is a working model of analysis and composition that says that each part is statistically *independent* of all other parts, much as the outcome of a coin toss is unrelated to the outcomes of other such tosses.[14] The parts are a set of independent random variables, to speak technically. We might put together or compose the parts much as bricks are stacked to build a wall, but now their orientations are random, unrelated to each other – not exactly bricklaying! Or, we might put together the parts into a sequence of moves, as in a game of Heads-or-Tails (HHTH HTHH . . .), each coin toss being independent of the others. By "independence" I mean that each part "knows" nothing of the other parts, so for example history and prospect have no other meaning than the cumulation of parts or steps. Only we outsiders (who can watch that cumulation) could have any sense of history; and as we shall see, this is a history that moves as the square root of time.

It is worth emphasizing how remarkable these are as a set of commitments. Were we to see a series of two dozen bricks that just lined up exactly – to choose a patently problematic case – it would be hard to believe that each brick's alignment was independent of that of the others, or at least of its neighbors. It would be more sensible, and surely more correct, to see such order as a sign of intentional device (someone deliberately lined them up), or at least of mutual local dependence, the orientation of each brick determining that of its neighbors. An account that still insists on the independence of parts or steps, as does random variation in evolutionary theory, bears the very great burden of not only showing how independence is sufficient for understanding what we see, but that independence is just about the only sensible story. Such a judgment of necessity is in part a matter of our aesthetics and conventions for such rules, in part a matter of our a priori expectations about the nature of the world. Is it really so random? One attraction of genetic accounts of evolution is that at the molecular-genetic level we find it much easier to believe in the ahistorical, nonintentional character of the cumulating steps.

What is most interesting about commitments to independence (or to any of the other strategies we shall describe) is the strength with which they are argued. So, for example, physicists employ great ingenuity in using independence to explain what appear as curiously orderly phenomena, such as freezing into a crystalline order or the development of permanent magnetization in an iron bar (or even the diffusion of a gas), so avoiding explicit alignment rules. One defends against alignment rules, or other heterodox interpretations of the rarer, quite orderly cases, so that the orthodoxy, here independence, is not only adequate to ordinary problems but is fully adequate to all situations. Twenty heads in a row (HHHHHHHHHHHHHHHHHHHH) – alignment par excellence – must be explained (and is explained!) in the same way that more varied and frequent distributions of heads and tails are explained. It would be subversive if twenty heads in a row were a matter of intentional device or unfair coins or whatever, rather than of independence.[15]

The crucial move here is a preemptive one – not uncommon in physical science, a move to rule some cases or problems as out of bounds, as more than extreme, as of no interest, so that the inadequacy of the orthodoxy in these arenas is of no consequence. To preserve independence as orthodoxy, some situations may be preemptively conceded by the

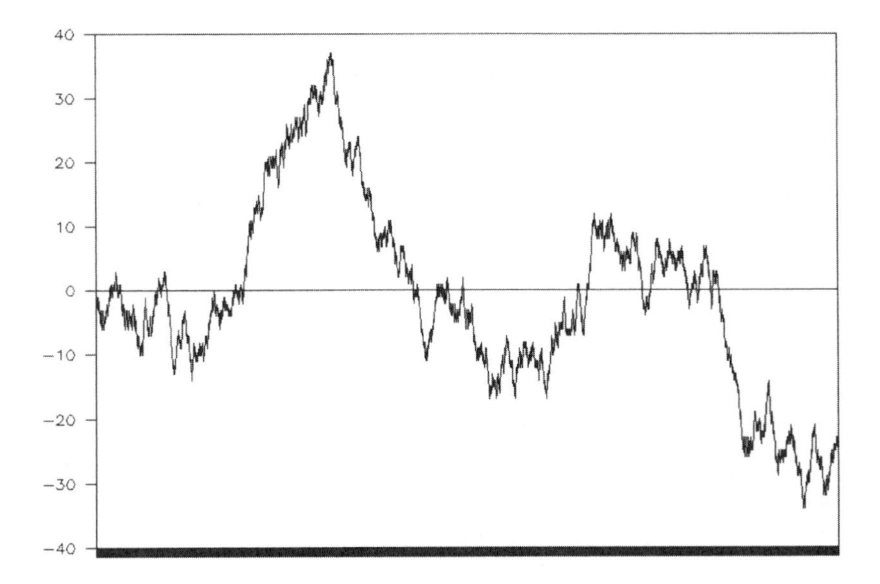

2.1. An example of a random walk, where UP is +1, and DOWN is −1

physicist to be a matter of bias, and of human intention and design and history, and so surely not matters of independence.[16] And if a physicist can so maintain a faith in independence, the analytical and compositional payoff of the strategy has proved to be substantial.

As I have indicated, the archetypal concrete model of independence is a sequence of coin tosses. Here, the degree of freedom of each toss is its outcome, a head or a tail. We may for these purposes ignore a multitude of other degrees of freedom, such as the exact position of the coin and its molecular constituents, the condition of the air around the coin, or the way the coin was tossed. And in ignoring these degrees of freedom, we get what we take to be a random toss, reflecting the strength of our ignorance of these detailed factors.

Now, for a physicist the outcome of any single toss is perhaps less significant than are properties of the sequence of tosses, expressed as cumulations and fluctuations. As for cumulation, consider the number of heads minus the number of tails. Conventionally one represents a head as +1, a tail as −1, and the sequence as a graph of the total value. (This is called a "random walk." For we might imagine ourselves moving

UP on a head, DOWN on a tail, and the cumulation would be our current position.) While the individual steps are unpredictable, the long-run average value is zero since the coin is presumed to be fair and so there will be roughly equal heads and tails. There are other, rather deeper features of the shape of the graph that are quite robust and predictable. For example, one can give probabilistic estimates of how long we might stay above the line before it crosses over. And, seemingly in contradiction – except that we speak of probabilities and not of single cases (the crucial move on the physicist's part) – one can show that the walk moves away from the x-axis, in the sense of its mean square distance: Average of distance-squared is proportional to time; namely, "distance" in this sense is proportional to the square root of time ("distance" $\approx \sqrt{t}$). (But, the average of all the actual distances (not the distance-squared), some of which are positive and some of which are negative, still is zero.) To do so, all we need assume is independence. In fact, the diffusion of a gas, which follows the same distance-squared rule, may be accounted for in terms of the presumed (but surely not completely) independent collisions of its component molecules.

Of course, it is just the mode of measurement, in terms of the mean or average of the distance-squared – what is called the variance – that is both the crucial physics and the right degree of freedom. The right degrees of freedom are not the outcome of each toss, not the obvious detailed property of each step or part, but the variance of the outcomes. Arithmetically adding up those parts' variances provides us with a good degree of freedom of the whole.[17]

(Technically, if our parts or steps are independent, the central limit theorem of statistics says that not only is the variance of a sum equal to the sum of the variances of its summands, but that is all that matters. Higher-order moments are in the limit unimportant; namely, independence means that the right degrees of freedom are variances.)

By roughly the same mode of argument, it has been shown that the future price of a stock may be accounted for in terms of the short-term, again presumably independent and so *un*accountable ups and downs of its price. (See Figure 2.1.) A stock's volatility (or the variance of its price changes) can be used to give a statistical prediction of its future price range: The price diffuses, so to speak.[18] No wonder securities firms might want to be sure that price movements are in fact independent, at least if they cannot reliably gain control over those prices or obtain reliable private information about those firms. In effect, despite our desire for an "edge," information about securities tends to be widely distributed, so that no trader has more information than is epitomized in the security's current price (so bids are not based on control or private information). And this for-lack-of-anything-better commitment to independence is all the more ironic when one realizes that the history of securities markets has been rife with shenanigans and insider information, with people who wanted to control or predict those short-term movements, thinking that such control was the only sure road to profit.

Returning to physics, what is most striking is the abstract manner in which physicists conceive of the world or model it, so it may be analyzable in terms of independent parts, so it may be taken concretely in terms of coin tosses or of picking balls out of urns. Perhaps the most powerful physical theory that starts off with the assumption of independence is statistical mechanics, the theory of the collective behavior of large numbers of similar objects. Classical mechanics assures us that the position of an atom at one time will be dependent on its position and velocity at a slightly earlier time. (Newton's laws may be expressed as differential equations.) But, if we choose as our parts not individual atoms, but large collections of atoms, as in a gas, then the complexity of those atoms' many mutual interactions might allow us to consider the detailed arrangement of the whole collection of atoms a short time from now (say after several collisions per atom) as more or less independent of the current arrangement.[19] A gas is said to be composed of independent parts, but now those parts are each of the various different detailed arrangements of the collection of atoms (a "state"), the totality of such parts being called an ensemble of states. Here, composition is about putting together a gas – whose atoms are, to be sure, constantly moving around – out of the very many different detailed arrangements of all of its atoms as a whole. (Analogously, a person is composed not of individual atoms or molecules, but of all of that person's aspects or ways of being.) And then the right degrees of freedom of a part are not those of each atom, but those of each arrangement or state, namely, properties of a collection of atoms. And the right degrees of freedom of the gas are actually averaged properties of the parts ("ensemble averages"), namely, properties of an ensemble of such arrangements – properties such as average energy, entropy, and volume.[20] Now, what is most attractive about the equilibrium state of a gas is that the actual values of these right degrees of freedom are quite insensitive to just which parts – just which arrangements or states – we happen to choose in evaluating those degrees of freedom.[21] In doing the averaging, properties of the parts, properties of an arrangement or state, are very likely to be properties of the whole: the energy of a state is very likely to be the average energy. Hence, what we see, literally a sequence of states from that ensemble, will have fairly unchanging values for the good degrees of freedom: The gas is in equilibrium.

For much the same reason – that many and varied independent parts seem to compose quite stable and robust wholes, such as a gas (or technically, that error or fluctuation only grows as $\sqrt{N}$, so the fractional error is $\sqrt{N}/N = 1/\sqrt{N}$, quite small if the number of parts, N, is large) – physicists will try to set things up so there is independence. They find and craft parts, steps, or building blocks that are as independent as can be made – so that they may employ the compositional technology of probability and statistics that goes along with independence. These technical formalisms are also employed because physicists cannot know, or do not know, the connection between these parts or steps or events. They are probabilists for lack of anything better. Similarly, insurance companies pool risks, and biostatisticians randomize trials of treatments, so that unknowable dependence is dominated by averages that presumably wash out such correlation, leaving a comparatively well-understood independence.[22]

To practitioners of a trade, commitments such as physicists' commitment to independence define what it means to be a member and to know how to go about doing your business. To outsiders, those commitments are most poignant and revealing. For practitioners, here physicists, go to very great lengths in reconceiving the world so that they can fulfill those commitments. In each of our subsequent examples, we shall see how other of the physicists' commitments function much like the one to independence.

DEPENDENCE, SPREADSHEETS, AND DIFFERENTIAL EQUATIONS

If parts are directly connected to each other one by one, they may be said to be *dependent,* the degrees of freedom of any part being systematically restricted by its neighbors. So the bricks of the last section might be aligned because each one was lined up parallel to the previous one. Or, think of a field as we describe it in chapter 1, in which properties or field values at one point are intimately dependent on properties of nearby points. (Between independence and dependence lies the Ising model (see the preface), where neighbors influence each other, but there is lots of otherwise random motion.)

We might even think to analyze a painting's composition in just these terms of dependence. Say the parts here are features of the picture (color, form, objects, . . .) and the degrees of freedom might be qualities of those features (hue, brightness, . . .) – each located at a particular point. It seems reasonable to expect that the degrees of freedom of one point must depend on the degrees of freedom of other points if the painting is to be seen as whole and coherent.[23] Imagine, however, not only that features are attached to particular points in the picture, but that the dependencies were only specified *locally,* relating features at one point to features quite nearby. A feature would affect features at distant points or affect the whole composition only by its propagated effect, from one nearby feature to another. Imagine, as well, that these relationships were given by a set of rules that were to be applied *uniformly* at every point, with no particular concern for larger context or global structure. One need not look at the painting as a whole to understand its composition; one need only look around at each point, and look with a uniform eye. Now I present this example because it seems so absurd. Composition in painting is ordinarily understood by looking at the whole of the work of art. The relevant analytic features or parts, and their mode of relationship, are to be understood in terms of that particular work and the tradition (say Christian painting) it participates in.

But, in fact, some stories of dependence are deliberately local and uniform, and everything is done to ensure that they can be seen that way.

Probably the major commitment of physicists is to dependence, locality, and uniformity. To emphasize again how curious this might appear, imagine a spreadsheet (as in *Excel*) in which the formula at every point or cell were of the identical form, and only referred to nearby cells.

In practice, the commitment to dependence, locality, and uniformity is a commitment to differential equations (which look like $F = m\, d^2x/dt^2$, or $\partial E_x/\partial x + \partial E_y/\partial y = 4\pi\rho$).* as a way of describing Nature.[24] At every point in space one expects the degrees of freedom of that point – the values of the variables at that point – to depend on the nearby neighbors' degrees of freedom through a generic rule that specifies how those values change over nearby distances: hence a *differential* equation. For example, it turns out that $\partial^2 T/\partial x^2 + \partial^2 T/\partial y^2 = 0$, where T or $T(x,y)$ is the temperature at a point, an equation that describes the steady-state flow of heat, says that the value of T at a point is just the average value of its neighbors' values – as in the spreadsheet in Figure 2.2. Hence, these physicist's commitments work in practice.

Now, in actual life situations, the physicist's commitment to uniformity cannot be absolute. Each locus or point turns out not to be the same. The actual world is rich with spatial inhomogeneity and peculiarity, such as particular charges located at particular points, or charges functioning as sources or sinks of a flow or a field. To tame such peculiarity, physicists separate the uniform or homogeneous aspects from the peculiar aspects, resulting in dependency rules that look like, in spreadsheet language: $c4 = (b4 + d4 + c3 + c5)/4 - Q$, where Q is a function of location (since the world is inhomogeneous), Q is $Q(c4)$. See Figure 2.2. (In differential equations language, this looks like: $\partial^2 T/\partial x^2 + \partial^2 T/\partial y^2 = Q(x, y)$, a uniform rule with a peculiar add-on.) This might well remind us of the division of labor, in chapter 1, between field, which is subject to uniform rules, and particles and sources, which are located

*For my purposes here, what should be noted about these equations and formulae, by the mathematically lay or the expert, is: (1) the patterns and symmetries in the formulae, much as we appreciate the symmetries in ornamentation or in works of art; and (2), the fact that the ∂ and d symbols mean that we should consider small changes in the quantity that follows them. Namely, these formulae are about symmetry and locality and differences.

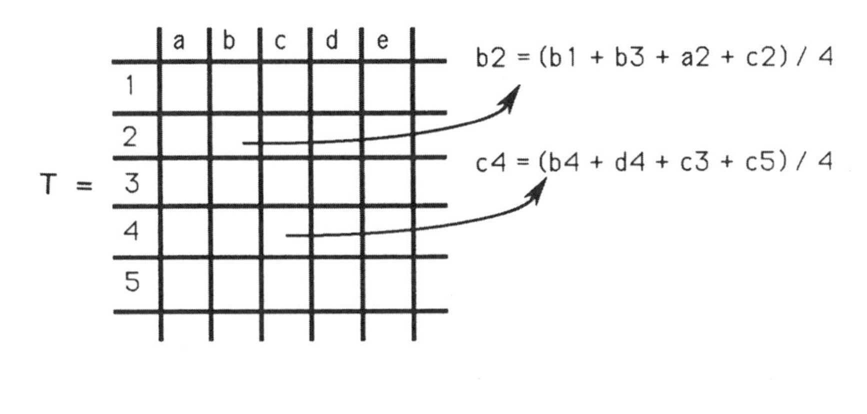

$$\equiv \frac{\partial^2 T}{\partial x^2} + \frac{\partial^2 T}{\partial y^2} = 0$$

2.2. A spreadsheet model of a field

inhomogeneously in space. Of course, it would be even more desirable if those peculiar features were not simply arbitrary, but somehow were given by the equations themselves – again locally and uniformly. So we spoke in chapter 1 both of wiring up the sources to the field and of the conservation of flow.

In each case, it is explicitly acknowledged that there may well be some unavoidable nonuniformity – which will be accounted for by the local particular feature, $Q(x, y)$, rather than by a change in the generic rule or differential equation. Remarkably, this strategy turns out to work in practice. It also works ideologically, since it allows for a commitment to dependence, locality, and uniformity, despite the world's peculiarity.[25]

Even if we were to have a local and uniform account of composition in terms of parts that depend on each other, we are presented with still another difficulty at the edges or boundaries of a system, when it would seem that the rules must be different. Rather, physicists describe what are called boundary conditions, such as "the temperature is constant on the surface"; and in so doing they are again separating off the uniform rules from the peculiarities. Even more desirably, imagine getting rid of the boundary – by extending it to infinity, or by folding the world back on to itself so that it has no boundary, for a two-dimensional world now becoming a sphere or a doughnut.

Practically, physicists insist on uniformity by avoiding or separating out arbitrary sources or ad hoc rules, or by reconstructing the world they are accounting for, say by folding a boundary back on to itself, so that the world cannot possess arbitrary features. Yet, as judicious scientists, they balance the desire for absolute uniformity and locality (what I take to be the demands of dependence) with the capacity to calculate and to make predictions. Allowing for arbitrary features or a boundary may make for a less aesthetically satisfying theory in ideal terms, but one that is more readily and practically and intuitively employed. (This judicious balancing is exactly the same sort of problem we faced in chapter 1 in discerning the appropriate division of labor in the manufacture of Nature.)

Even if there is locality and uniformity, again there may well be mixture of dependence and independence. In a magnet the atomic spins, as microscopic magnets, will be pointing randomly due to thermal motion. But the spins' mutual interaction would make them align with each other.

Given the commitment to locality and uniformity, there is still further room for maneuver, extending our notions not only of uniformity but of locality. What does it mean to be local or nearby? In everyday parlance we know exactly what we mean – namely, close by in space, or just before or after in time, or perhaps connected by a direct line. Physicists extend everyday notions of nearbyness so that they can continue to use the dependence mode of analysis and composition. For example, if distance and closeness might be specified only along a particular kind of path, then physicists can reasonably revise their notions of nearbyness. If paths of connection are restricted to a plane or a line, formerly nearby points that are now not in that plane or on that line are now not considered nearby. Conversely, closeness, or at least connectedness, is extended by allowing for more paths between two points, such as diagonal lines, for now more points are nearby and influential upon each other. Given the usual kind of regular grid: on a line, a point has two neighbors; on a square grid, it has four neighbors; and on a square and crosshatched grid, it has eight neighbors. Now a line is one-dimensional and a plane is two-dimensional. So we might expect that with higher dimensionality, points will be better connected and thereby more nearby to each other, and so it will be harder to shield them from each other, recalling our concerns in chapter 1.[26] Given a commitment to locality, the dimensionality of space is a deep feature of the actual physics of the world. Dimensionality, taken as connectivity and nearbyness, is a good degree of freedom and is said to be dynamical.

When physicists talk about four-dimensional spacetime in special relativity, and then in quantum mechanics and modern particle theory, they are saying that there are more connections between points or events or properties, greater influence, less freedom for any single one, than if they could consider just three spatial dimensions, while time played a separate "tick-tock" role, merely marking events with their time of occurrence. Points are in effect closer, more nearby; events are influential on each other (at least if they have what is called a causal or "timelike" connection); properties are mutually constraining. Even more generally, if in fact more influences are clearly operative than are allowed by the conventionally nearby points, what physicists do is to extend the number of dimensions of the world – namely, the paths of direct connection – so that all the relevant influences now really are nearby. And, it should be noted, those dimensions are not only spatio-temporal, but refer to other of the system's degrees of freedom such as electrical charge or isotopic spin. Conversely, we might decide that some connections are in fact not so intimate, and so reduce the dimensionality of space. High-temperature superconductivity seems to require that we consider current flow in planes of matter, namely it is two-dimensional and not three.

In sum, physicists' commitment to locality, or nearbyness of influence, may encourage them to enlarge the world they analytically consider, namely, to increase its dimensionality. And those dimensions then delineate just the right degrees of freedom. What is to count in the composition of the world will now be nearby enough to each point to make up the world according to uniform rules. Moreover, perhaps those rules are not only the same everywhere, but they look the same for many of the dimensions or degrees of freedom. The rules are then local, uniform, and *symmetric*. For example, the Dirac equation for the relativistic motion of an electron, $-(ih/2\pi)(\gamma^x\partial\psi/\partial x+\gamma^y\partial\psi/\partial y+\gamma^z\partial\psi/\partial z+\gamma^t\partial\psi/\partial t) + mc\psi = 0$, looks the same for all the dimensions: x, y, z, and t.

Now my point here is not so much to discuss particular mathematical models as to indicate the strategies employed once we are committed to dependence as physicists have come to understand it. The justifications physicists offer for why dependence is a good and successful strategy are: First, as we discussed at the end of chapter 1, the physical world is characterized by particles and fields that are wired up nicely to each other (as in what are called "gauge theories"), in an efficient division of labor, so to speak; and hence spacetime could be taken as uniform or homogeneous except for source particles, and hence uniform rules might well work. And, second, that special relativity's incorporating of space-time symmetry requires locality – namely and technically, since the basic dynamical physical laws are differentials or local in time, so by symmetry they must be local in space, too. And so the physical world, as physicists discover it and conceive of it, encourages their commitment to dependence – and to uniformity, locality, and symmetry as the general features of dependence.

The technology of differential equations is attractive because by its conventional practices it automatically incorporates the strategy of dependence and its features. The differential equations physicists canonically employ express local dependence that is uniform and as symmetric as Nature allows. Moreover, the fact that differential equations in general build in the assumption that for the most part the world is continuous and smooth (namely, the actual smallness of those marginal or differential parts) affirms one of the great traditional metaphysical commitments.[27] And that one can get away with uniformity and locality and continuity – in an automatic and natural conventional way, adding in whatever source terms, boundary conditions, and choices of dimension one needs – is taken by physicists as confirmation of the basic strategy.

Recall that coin tosses are the archetypal model of independent parts, and Nature is then modeled as being a consequence of a sequence of such tosses. Here, for dependence, the model parts are the loci of an interconnected matrix, such as the field values at each point; and in satisfying their interconnectedness (in a spreadsheet or in a differential equation) Nature is fulfilled. Now if physicists choose the right parts, not only will the rules of dependence or interconnection be tractable, those parts themselves will prove to be readily isolated from each other and individual (each possessing good degrees of freedom) despite their mutual dependence.[28] We may speak of the value of a field at a point: the velocity of flow of a fluid or the electrical field, each at a point. Parts may be considered literally individually.

Similarly, we may recognize separate individuals in a liberal democracy or an economic market, despite the individuals' mutual interaction. And again, in a liberal democracy and in markets, uniformity and locality are instantiated: uniformity being that the same rules are presumed to apply to everyone, and locality being that individuals make their decisions themselves or through interactions with their neighbors, each individual doing the best he or she can.[29]

There remains a curious irony. The strategy of independence, in the limit, may lead to a strategy of dependence (albeit at different scales, and the change of scale is crucial). Now, a gas literally diffuses – that is just what it does. Recall that a random walk of independent steps might also be said to diffuse, as we might expect if we think of a gas as being composed of randomly colliding molecules. Yet actual diffusion is usually treated as a matter of dependence, namely, the differential equation for diffusion (in fact, just that differential equation that says that the density of molecules at a point is the average of the density in the neighboring area). Those more or less independent steps or collisions embody a set of local uniform requirements – that in the long run flow is downhill and that in a steady flow there is neither a net gain nor loss of material at any point, unless there is a source or sink there. And this is also what is expressed in the differential equation for diffusion. The trick here is that these requirements are enforced not on individual steps (which remain independent) but rather on very large collections of these steps. (Technically, what is crucial here is that micro-fluctuations are comparatively unimportant since the number of molecules is very large – the scale change from microscopic to macroscopic doing just this work.) Nature forces the physicist to recognize an intimate connection between two very different strategies; and physicists then celebrate this fact, namely, the atomic account of the macroscopic phenomenon of diffusion.[30]

ADDITIVITY AND THE CALCULUS

I have suggested that composition is the putting together of components or the cumulative effect of marginal influences. Sometimes this putting together or cumulation may be literally arithmetic, the addition of numbers.[31] Something is composed or put together by literally summing its parts. Our intuition about such an accounting is that: (1) no matter how we do the addition, how we group the summands, the answer should be the same, and so there is just one answer – what I shall call path invariance; (2) adding on small parts will have small effects – marginalism; (3) something big may be analyzed into smaller parts, parts that when added up compose that big thing (or number) – invertibility or decomposability; and (4) like segments on a ruler, the parts can be lined up and so summed – a kind of linearity.[32]

Additivity is such an obvious strategy that we might think it easily applied everywhere. All we need do is line up the parts, or count things up, whether they be billiard balls or atoms or whatever.[33] (But note that we are not putting together the actual balls or atoms; we are putting together, actually adding, measures of their size or number or other such properties.) It should not be difficult to achieve path invariance, marginalism, invertibility, and linearity. But Nature proves not always so tractable. For example, if in combining two parts there is substantial interaction, as in a chemical reaction, things do not add up so straightforwardly.[34] To save the appearance of additivity, physicists employ a variety of devices.

Perhaps the objects do not line up on a ruler. So we invent vectors, in which the addition along each dimension is linear.[35] But perhaps the order in which we add those vectors matters (what might be called a path dependence), as is surely the case in adding up vectors on a sphere. Near or even at the geographic North Pole, going three miles south and two miles west is different in its consequence than going two miles west and then three miles south. So geographers, and scientists more generally, have developed correction rules that keep track of the consequences of different orderings. Physicists call them commutation relations. They can then properly or "canonically" perform any particular addition, taking into account effects due strictly to the path that was followed, a theme we have already encountered in chapter 1 when we discussed fields.

Historically, perhaps the most serious challenge to additivity came when the parts to be added were segments of a curved line.[36] If a line is curved and smooth, how is one to add up its parts so as to find out how long it is? Does the length we find depend on how we break up the line into those parts or pieces? Our intuition tells us it had better not. The great invention here was the integral and differential calculus, which provided among other of its innovations a canonical way of converting such curved lines into marginal parts (velocities or "tangent vectors" or derivatives) and a way of adding things up (the definite integral).

But if the calculus is to make the changing and curved world additive, it demands that the world be smooth.[37] Then its process for finding nice marginal parts, the derivative, a process which produces nicely summable objects, is legitimate and actually works. There is path invariance, marginalism, invertibility, and linearity. The calculus is such a powerful device for analysis and composition that physicists are tempted to assume that the world is smoothly changing or can be made so – say by averaging or smoothing, or by isolating its non-smooth places, or by finding a perspective from which discontinuities may be avoided.

In part influenced by the culture of the physical sciences, neoclassical economics often relies on a "Principle of Continuity," for example as defined by Alfred Marshall in his *Principles of Economics* (1890), to assure the possibility of both analysis and composition through the calculus "that there is a continuous gradation" in preferences and time periods, and "that our observations of nature, . . . relate not so much to aggregate quantities, as to increments of quantities."[38] Differences in kind are to be made into differences in degree, for then the calculus and its apparatus and language applies in full force.

Physicists feel that they have the right degrees of freedom if the parts or their properties can somehow be added up, correction factors and all, just as ordinary numbers are added up, willy-nilly, in any order, with any grouping. If the world can be made to go this additive way, then the partitioning does lead to unique sums; the parts have marginal effects, they will recompose something through addition, and they can be lined up. Now if the sum is still unavoidably sensitive to the particular path along which we do the addition, physicists may search for the "minimal path," say the path that takes the shortest time, as a means for distinguishing one sum as the correct one. Or, they average over all paths, all the ways of doing the addition, and say that that average is the proper notion of a sum. In any case, by device physicists endeavor to find just one canonical way of choosing and adding up the parts, thereby getting unique sums. As long as they are acting within a tradition of such devices, physicists believe that whatever one needs to assume to achieve that additivity might somehow be right and true and be of deeper significance. It turns out that correction factors indicate that the world is in some sense curved; minimal path rules point to conservation laws; and averaging suggests that the world is plenitudinous, all possibilities playing a necessary role. Seemingly technical devices point to deeper features of the physical world, a phenomenon we shall return to in chapters 4 and 6.

More generally, whether it be a matter of independence, dependence, or additivity, physicists design and then redesign the parts so that they will act right, each part being a kind of individual that is in accord with informal conventions that have often worked before. And then they invent mechanisms and setups so that the parts in concert will do the compositional work that needs to be done. And those mechanisms point to pervasive characteristics of Nature as physicists understand it.

DISJOINT FUNCTIONALITY AND INTERPRETABILITY: BUREAUCRACY, FLOW PROCESSING PLANTS, AND OBJECT-ORIENTED PROGRAMMING

One of the great advances in the study of human physiology was the realization that the body is to be accounted for in terms of system, flow, and process. As we shall see, system, flow, and process provide for very different sorts of degrees of freedom and handles than do the properties of a conventional, individual, particulate atomic unit or even a spatial field and its value at each point (whether independent, dependent, or additive). Here, the phenomena of interest are the *paths* of flow, whether they be represented by means of pipes and of decision and control elements (acting as valves, switches, and interchanges), or by global patterns of flow and congestion. These latter flow patterns may even be conceived of in terms of larger-scale flows articulating into smaller-scale flows – a story that might be called top-down – much as eddies of smaller size are present within larger-sized eddies, or bodily systems are articulated into subsystems. Or flows, especially if they are taken as matters of pipes and valves, may be built, bottom-up, out of smaller-scale flows, a Tinkertoy story. Here I shall concentrate on several simpler models of motions, flow, and process: namely, bureaucracy, flow processing plants, and "object-oriented" computer programming.[39]

Surely, the bureaus that make up a bureaucracy, or the vats that make up a chemical processing plant, or the stations of the mass production factory are to some extent independent of each other.[40] Yet the flows among them depend on each other, and sometimes we might even think of the process as a marginally additive one whose sum is quite dependent on the path we follow. Still, these kinds of partitionings often do not prove to be the right ones if we are concerned with flow and process and path. The right parts of a bureaucracy often prove to be its bureaus, taken as active or reactive agents, and their processing rules. And the bureaus' distinctive features are what I shall call *disjoint functionality* and *interpretability*.

Before going further, I should note that the applicants to the bureaus are treated here as external factors, their relevant properties or degrees of freedom determined by how we have set up the bureaucracy. Clearly, the bureaucracy will behave differently depending on that external environment. Of course, we might treat the bureaucracy as external, the flow of actors or data as fundamental, as we shall see shortly. What is crucial is the separation into internal and external (or the separation of uniform field rules from arbitrary sources, as we discussed earlier). With such a separation the physicist can go to work, fruitfully it turns out.

The bureaus of an idealized bureaucracy are *disjoint functional* units, each performing a well-defined processing task, so that it may ignore other bureaus unless it needs their specific help (or its help is needed by one of them). Bureaus have specific entrances and exits, and inputs and outputs, yet often we do not know much of their inner workings. Those workings are deliberately shielded from our eyes. Bureaus may be said to be black boxes, inside of which they do something to their applicants or inputs, and then send them off. Now if we do have a chance to examine any single bureau more closely, penetrating the walls of the black box, we may find that its work is done by a set of disjoint functional sub-bureaus (and these may then lead to sub-sub-bureaus) – something like nested Chinese boxes. The task of any part, any single bureau, may be *interpreted* into a linked set of parts.[41] Such articulation may lead to reticulation, a net-like or tree-like system. Often, there is a hierarchy of levels, possibly crossed over. The actual path through a highly interpreted system might even be unending and recurrent, individuals stuck in the system never to emerge.

Each bureau not only does something to each applicant and sends the applicant on its way, but it may even have no memory of or detailed information on an applicant once the applicant has left the bureau's charge. Applicants may carry their own records – think of patients in a hospital – and they are fully known at any time by their own records. Where an applicant goes next is determined by its properties, or what might be called its needs, and by the available bureaucratic resources – not by any single prescribed path within the bureaucracy. So the flow through the bureaucracy cannot be conveniently epitomized by predetermined linkages among the bureaus. Like a road map, the organizational chart does not say which path will be followed. And the travelers or applicants are often not so constrained, and the congestion they experience not so insignificant, that we might say that they travel the quickest path, or follow another such principle – although it is a recurring dream to find such a principle, a principle that would prescribe just one path for each applicant. Rather, the flow is determined, at each moment, by the applicant's current properties or record and the available resources. We might say that the applicant triggers the operation of a particular bureau, and a path is a sequence of triggerings, each triggering a function of the current properties and availabilities.

Again, in general it is not at all clear that something that enters the system will emerge. Now, if the bureaucracy were like a chemical processing plant or a mass production line, each of which provides a comparatively fixed path with few options, then we know what should happen at each stage; and even if there is feedback or sendbacks (for fixing), we believe that most of what comes in must leave in finite time. More arbitrarily, if we have a disjointly functional and highly interpreted bureaucracy, with no fixed path, with applicants or objects calling on bureaus by their own rule-driven determination (what is sometimes called an object-oriented program), then an applicant might well be expected to have an idiosyncratic path or history.[42] For example, there will be congestion at bureaus which have to serve more than one applicant at one time, and so similar applicants might have different histories due to conflicting demands for pathway.

Empirically, it turns out that most of the time there are dominating patterns of flow. Of the many possible patterns, comparatively few are ever realized. There are perhaps two sorts of pattern: Either all flow lines are uniformly populated, with small and random fluctuations, or there are genuine highways and archetypal routes. (These sorts of patternings give physicists hope they might deduce the dominant patterns from the general principles I mentioned above – either about plenitude or minimization.)

Another model of disjoint functionality and interpretability is a telephone switching system. Telephone numbers are peeled off, numeral by numeral, and set off a hierarchy of switches to create a path between two points – first area code, then central station, then specific phone line. There are multiple paths in the telephone system; and redundancy in the paths permits connections which otherwise have conflicting needs for pathway. Of course there are calling patterns, and one can design the switching and cable system so that overload is avoided most of the time. Still, besides occasional overload, some of the time there will be breakdowns and errors in switching. One needs to design the telephone system so that it is robust enough to compensate for at least minor breakdowns and is capable of correcting or at least recovering from some errors.

Now the strategy of bureaucracy and articulated flow, and what I shall in the next section call procedural analysis, is nowadays much more common in the design of actual experimental equipment, and in computation, engineering, and organismic biology than it has been so far in theoretical physics.[43] Rarely must the world go this way for physicists, when they explain physical phenomena. But when physicists take on complex systems, or when they advise on war planning and the economy, they often do conceive of those systems in just these bureaucratic terms.

These are, to be sure, remarkable abstractions, whether the parts and degrees of freedom are bureaus, rules, and kinds of applicants, or are global patterns – but perhaps no more remarkable as abstractions than are parts taken as independent, dependent, or additive. And it is just that level of abstraction, exemplified in so concrete a model, that is so characteristic of how the world must go for many a physicist.

Now we still have to assure ourselves that these parts, in concert, will compose Nature. But here physicists rarely have some nice analytic mathematics to do that work, even for slightly realistic cases. So, alternatively, one sets up a simulation or a game – invoking the rules and patterns for a sample of objects – and compares that simulation to actual paths.[44] Along the way, we may well discover that a particular simulation is subject to idiosyncratic behavior or infinite regress, or is quite sensitive to error and to the breakdown of a part, even if actual Nature is rather more convergent and robust and patterned. What physicists characteristically do is to restrict the environment, the kinds of applicants or objects, so that the applicants' demands are now likely to be readily handled by the model system. Although we would at first appear to have generic parts and a general-purpose system, in fact the strategy of bureaucracy only works well as a partitioning if extreme cases are given special treatment. The physicist has to get the degrees of freedom right and also properly identify the extreme cases (those that need not be accounted for).[45] In actual bureaucracies there is much backstage dealing and friendly exchange, so as to take care of exceptional cases and to maintain the front-stage image of disjointedness and interpretability. And in telephone systems and computer programs and processing plants, careful restrictions on inputs, and ad hoc but quite workable stopping rules, maintain convergence. So, in fact, the physicists' moves are in effect justified.

The remarkable feature of bureaucratic analysis is that flows become discretized into a lexicon of steps or bureaus or patterns, each particular history sharing in that set of decisions or patterns. And the degrees of freedom are, at least initially, features of the bureaus and their applicants or features of the patterns of flow. Interpretability, like independence, dependence, and additivity, is both recognizably an abstraction and an effective mode of analysis, so making individuals into parts.

SEQUENCE AND PROCEDURE

But what goes on inside one of those bureaus or vats or processors or switches or stations? There might be sub-bureaus and sub-sub-bureaus, a matter of disjoint functionality and interpretability – all the way down. But eventually, at least in actual practice, we come to some primitive analytic element, say someone doing his or her work, performing a set of procedures.[46] Bureaucracy then yields to procedure.

The crucial feature of procedural modes of analysis and composition is *sequence* (rather than independence, dependence, additivity, or disjointness and interpretability). What is done is done in an order, and there is precedence and subsequence. So if the procedure has been properly set up, we are in effect guaranteed that at each stage we are ready for what comes next. Imagine budding up a solid crystal, from a liquid, atom by atom now accreting to the solid. Or consider driving instructions, which are often such sequential procedures. Each instruction makes sense if we have followed the previous ones, although we usually try to build in some robustness so that there is an easy recovery from minor error ("You have gone too far if . . .").

Or, consider the problem of pruning a tree of connections to get rid of branches that go nowhere. The following procedure will do some of the work automatically and also locally: Remove any branch that is only connected at one end and not connected to the ground or an edge; repeat this process until no more pruning occurs. Each step sets one up for the next step. So a long stray branch will be pruned piece by piece.

For physicists, procedural analysis is usually a mode of actually solving a realistic problem. Characteristically, procedural analysis converts a static problem or goal into a sequence of steps, to be performed one after the other, as in a program or an algorithm.[47] An optimization principle, say that light travels in the least-time path, becomes a differential equation which prescribes the next step in a path given its current endpoint. Now, of course, differential equations are surely accounts of dependence. But when physicists are concerned with solving a concrete problem, they must often solve those equations approximately and numerically. And then a differential equation might be considered a schematic procedure, about one step being literally added on at a time.

More generally, approximative methods for solving physical problems often involve procedures, which procedures are then shown to have deeper physical meaning than we might expect from a technical method. For example, consider Monte Carlo methods, which are procedures that involve a sequence of random trials. They may be employed (using random walks) to solve the differential equation that describes the flow of heat. We might try to argue that those methods work because molecular collisions are the means by which heat flows from one point to another, and so the Monte Carlo method simulates actual physical processes. Those methods very differently employed are as well good methods for finding an efficient layout of a computer chip. And they also may be used to simulate the process whereby we anneal a metal to get it into a more defect-free form. In effect, one becomes more efficient and less wasteful.

In each case the parts are steps within a procedure. Those marginal changes, which could be about dependence, are just what we do next. Those random trials are each a prescribed sequence of choices, each choice determined by the toss of a die. In general, the parts might well be independent objects, or dependent field values, or demographic particles, or bureaus. But what distinguishes procedural parts is their being steps in a sequence of precedence and subsequence.

As in bureaucracy, procedures can employ other procedures, including repeatedly employing themselves. Algorithms, computer programs, and operational plans are examples of analysis that represents composition as a sequence of steps. As in all accounts of analysis and composition, one has to convince oneself that the combination of parts, here the sequence of steps, is in an acceptable sense the same as what it is supposed to represent. Does the procedure adequately represent the world and what is supposed to happen? As in bureaucracy, complex sequences may have ranges of consequence that are not at all apparent, their being an adequate analysis of the problem perhaps not so obvious. Again, physicists simulate their workings (set them up and play them out, as in a game) and see if they really do the compositional work. No matter how sure we might be about the procedure's structure or logic, we verify a computer program by actually running it on a suitable sample of data, seeing if the results feel right.

As we have recurrently seen, the physicist's obsession with proper partitioning becomes a technical problem in pursuing a particular strategy. Here, one obsession is with sequence and its switch points – what might be called a top-down analysis. And the physicist's other obsession is with endowing each step or part with the right properties, and then composing those parts into subassemblies or procedures, again with the right properties – a bottom-up process. The computer scientist Alan Perlis describes the dialectic of top-down and bottom-up, of what might be called analysis versus composition: "[The computer language] PASCAL is for building pyramids – imposing, breathtaking, static structures built by armies pushing heavy blocks into place. LISP is for building organisms – imposing, breathtaking, dynamic structures built by squads fitting fluctuating myriads of simpler organisms into place."[48] There is an enormous stylistic difference between top-down analysis and bottom-up composition – both being modes of partitioning – even if they were to prove to be logically identical.

In procedural analysis, the right parts and their degrees of freedom are the modes of sequence and control and the steps themselves. Now, if we change the sequence of the steps in a procedure, the procedure – if it works at all – will do very different things than before. The effect of a change is almost never marginal. And if we change a step within a procedure, the procedure is again likely to change radically. So one might imagine a desire to find a procedural analysis which is robust – changes in the sequence or the steps have little or no effect on how the procedure works. Or, we might want a procedural analysis that is canonical – any change in a sequence or a step will have dramatic effects. Recalling the division of labor in chapter 1, so-called work teams in factories are comparatively robust and marginal, while assembly lines are canonical.[49]

To close these considerations of partitioning strategies, I want to note that parts are multiply meaningful and that the choice of strategy is not nicely given by Nature. First, we often find that a part may do different kinds of compositional work at the same time. A particular atom randomly bouncing around off of other atoms is also a part of a diffusing gas as well as a marginal part in the ratios of the gas's chemical components, components which may well react with each other. Physicists take this multiplicity as being not only about different ways in which they might conceptualize the world; it is also about how the different ways are intrinsically connected, that is, about the physical world itself. Chemical reactions take place through the interactions of colliding atoms, the rate of reaction being in part a matter of the rate of random collisions. Second, each way, each strategy of partitioning, brings along with itself a tradition of conventional, successful practices, artisanal skills so to speak. And those traditions may well differ substantially among the modes of partitioning. To choose a strategy of partitioning is not only a physical choice; another strategy might work well, too, if differently. It is also a cultural and professional choice, about what kinds of explanation this physicist seeks and what kinds of skills this physicist possesses and may apply to this problem. Put differently, it is an economic choice about the kinds of products you will make so as to capitalize upon your comparative advantage over others.

III

PARTS ARE COMMITMENTS

Each strategy of demonstrating how the world is made up of parts brings along with it a practical aesthetic: archetypal models, and conventional modes of understanding those models, that help ensure that the analytical technology that goes along with that way of partitioning actually does do what it is supposed to do. If we hew closely to some model, such as a random walk or a particular differential equation or a clockworks, we are more likely to do an analysis that will in the end really lead to composition – at least for a range of problems. Independence of parts is often expressed in terms of the technology of stochastic processes (randomness); dependence is taken in terms of a spreadsheet or a differential equation (locality and uniformity); additivity is expressed by the calculus (path invariance, marginalism, invertibility, and linearity); disjoint functionality and interpretability is bureaucracy or object-oriented programming; and sequentiality is procedural analysis. If the parts' desiderata are not met in a situation, one diddles with the world and redefines the requirements (such as locality) so that they are fulfilled.[50] Then the physicist can go to work: Individuals are now playing the right sort of parts and physicists are likely to have some degrees of freedom they won't be sorry about.

I want to reiterate that to be made up of parts is to be made up of a particular kind of parts in a specific fashion. It is not enough to know the calculus; it is crucial to know how to use it in a specific arena, applied to particular problems. I have described each of these strategies of partitioning the world in some detail, because it is in the details, in the particular applications of the archetypal models, in the conventional ways of doing each partitioning, that lies physicists' actual practice and commitment. There are perhaps many ways of exhibiting dependence and naming its characteristic features. But only a very small number of ways are actually employed in practice. So not only is dependence often about differential equations, it is mostly about a remarkably small variety of them – which physicists and mathematicians may then justify as the only good or "well-posed" ones.

I have been describing just how physicists go about fulfilling their commitment to analysis and composition, how they deliberately and obsessively pursue such commitments, insisting on partitionings of particular sorts. (In chapters 4 and 5 we discuss how physicists justify the natural world's allowing them to do so.) Such an obsession is rational, in that there are large arenas in which it is productive. Partitioning does give physicists good handles onto the world. Giving an account of how something is put together leads to ways of getting hold of it: by manipulating its parts, by interfering with the process of composition, and by employing degrees of freedom of the whole thing that are seen to derive from its mode of composition.

Whether it be small marginal changes that physicists add up, or components that they fit together, in the end what physicists are often insisting upon is that big things are made up of littler ones. Accounts of analysis and composition show, along the way, what physicists mean by big and little and by simple parts, and what they take as obvious or automatic ways of putting things together. Such accounts also show how far physicists are willing to go in treating an actual object abstractly, so that they may see an identity between that object and an account of its composition. Ice or a metal bar are not obviously a set of atoms linked by random interactions, nor are they obviously "massive continua" whose characteristic motions are vibratory waves (sound). But for physicists each of these are fundamental partitioning abstractions employed in accounting for the behaviors of these solids.

More generally, we appreciate the meaning of our culture's demands by watching just how we go about insisting on them in practice. The peculiarity of demanding independence, dependence, additivity, disjointness, or sequentiality is striking – especially when we consider that what we often do in our everyday nontechnical lives is to violate these demands and go around their strictures (those backstage conversations in a bureaucracy, for example), so as to make things work reasonably smoothly.

These first two chapters describe how Nature is a factory of laborers and a mechanism of parts. The physicist's problem – and thus the motivation for the obsessions and commitments – is to figure out how the labor is divided and coordinated, or how to define both the individuals as parts and their mode of composition. Nature seems amenable to fairly sharp divisions and partitions. Yet in practice there is some slack. There is room for realignment and reorganization and readjustment. Physicists may well say: Nature *must* go this way – so that a metaphor is authoritative or an analogy is destiny. In practice, they mean that Nature can be made to go this way, for there is more than enough room for the needed acts of interpretation. In the next chapter the informing metaphor is marriage and interaction, rather than division or partition – namely, an account of exchange and the extent of the market, to recall Adam Smith on the division of labor. Here we shall have a story of fruitfulness and of how possibility becomes actuality. But again, what will be most interesting is how physicists interpret marriage and kinship as saying how the world must go.

More generally, we have described in these first two chapters an obsession with getting the world in order, with cleaning up the world: Particles are to be well defined, with no leftover properties lying around. But since there will be leftovers, fields are to be well connected to particles, and those fields are to be tamed by conservation laws. And then, everything must add up, again with no leftovers. Alternatively, what we take as whole and complex is inevitably a neat sum or fitting together of nice clean parts or the product of a sequence of simple operations. We can divide and partition and so conquer a world. Here is a story of order and discipline.

But there is another story. Rather than being neat and clean and fully accounted for, why not let happen anything that can be imagined, allow any combination that is not explicitly forbidden? The obsession with control leads, by inversion, to a world where anything that can happen will happen. Yet, as we shall see, the commitment to order and lawful mixture remains.

THREE

Freedom and Necessity: Family and Kinship

Recapitulation and Prospect; Kinship, Exchange, and Plenitude; Systematics in the Field; The Problem of "Quite Rarely"; Markets and Fetishes; Taking the Rules Seriously; Structure and System.

THE ARGUMENT IS: THERE IS A REMARKABLE ANALOGY BETWEEN kinship systems, particle physics, chemistry, and market economies. All may be accounted for by stories of fair exchange: of women, elementary particles, electrons, and currency and goods, respectively. Actual social systems and Nature are taken as the consequence of the necessary occurrence of all exchanges that are not forbidden ("plenitude") and also the fact that names or labels on objects fully characterize them, both in their interaction with other objects and in their classification into groups of like objects ("fetishism" and "nondegeneracy"). Once we take on the world as a system of exchange, it seems we are also committed not only to plenitude and nondegeneracy, but also to rules of exchange that balance or conserve the flows of exchanged objects, the set of exchanges being the glue which ties the system together – at least if we are physicists. Here, the degrees of freedom are those nondegenerate fetishized labels (particle properties, for example) interpreted by the rules of forbiddenness, rules which reflect the orderlinesses or symmetries of the system. And the conservation of flow (say in balancing goods and money, or charge, or energy that is exchanged) also implies that we treat the exchanged materials as fully commensurable with each other.

The systemic structures that arise as a consequence of the processes of exchange – such as markets, conventional societies, and the "families" of elementary particles – are not at all obvious were we told about just those processes. The structures are surprises and are said to be emergent. Whether it be in evolution or quantum mechanics, those surprises are, however, most often known already ahead of time; for the properties and rules are actually designed to give an account of those structures.

"Who ordered that?"

– I. I. RABI, referring to the muon particle.[1]

"Freedom is the recognition of necessity."

– SPINOZA[2]

RECAPITULATION AND PROSPECT

In the first chapter, our story concerned the division of labor within a factory. There we noted the intimate connection between the design of a factory and of the laborers who do the work within it, what has been called scientific management. Walls, particles, and fields work together to manufacture Nature, the degrees of freedom distributed among them. In the second chapter, a controlling metaphor was the clockworks mechanism, just that mechanism that Maxwell revered in the passage we quoted in the preface. Nature was made up of individuals taken as parts, parts that fit together. And choosing the right parts leads to getting the degrees of freedom right, although those degrees of freedom might not belong to any single part. In this chapter, we shall be concerned with how those individuals interact, how the processes of exchange and the extent of the market (to recall Adam Smith) produce Nature.

Compared to the factory and the clockworks, the institution of the family seems primitive. We are sometimes told that taboo, incest, and kinship rules are atavisms. Yet, it turns out that within this story of marriage and fetishism* lies much of modern market economy, biology,

**Fetishism* isolates some thing from its larger context of interaction, treats it in terms of properties taken as belonging to it, and takes those properties and their associated powers or potencies as intrinsic to that thing's nature. Hence, seemingly inanimate objects are taken to possess magical properties and powers (presumably reflecting their larger context); and parts of a more complex thing surrogate for the whole thing, as in the rhetorical trope of synecdoche.

chemistry, and particle physics. For families may be taken as kinship systems, as systems of relationship and exchange. Systems of that sort also model an economy or the interactions of elementary particles. The degrees of freedom within such systems are the names that designate our relationship to others, such as kinds of cousinhood or prices or chemical valence.

I. I. Rabi was one of the great physicists of the twentieth century. Rabi's comment was made upon the identification of the muon particle in 1947, a particle that was, as far as could be told, just like the electron except that it was about 207 times heavier. The muon did not seem to have a distinctive name (that is, a set of sufficiently distinctive properties other than its mass). It was a Doppelgänger: it was too much like the electron. In fact, it should have been "degenerate with" the electron, having the same mass.[3] And as we shall see, from the physicist's point of view there was no good reason for the muon to exist. It was not required that there be this heavy electron in the known system of Nature – yet there it was. For while physicists abhor degeneracy (here, in mass), they are equally upset by mass differences that are not accounted for in terms of properties or degrees of freedom.

There should not have been a muon. But since the muon was found, physicists are forced – forced, as we shall see, by their commitment to the model of kinship and exchange – to ask: Why does the muon exist? What distinctive degree of freedom do it and the electron differently instantiate? How is the muon necessary? Eventually, physicists redefined their notions of family structure so that the muon did become necessary. And, in fact, they ended up defining three families: that of the electron; that of the muon; and the family of an even heavier electron, the tau (discovered in 1975).[4] In this physical world, if signs of new degrees of freedom appear, such as the muon, our notions of necessity (here, the kinds of families) need to be revised. For what is necessary is just the source of possibility and freedom.

KINSHIP, EXCHANGE, AND PLENITUDE[5]

When anthropologists have set out to understand kinship structures – namely, who might be married to whom in a society – some of their most crucial realizations have been that: (1) marriage is a matter of exchange, the exchange of women; and (2) if such an exchange were lawful and so allowed, then such a marriage would actually be observed, that is, it would eventually take place. Not any woman could be exchanged, and so married; and only certain arrangements or exchanges among families are thought to be fair and balanced. The rules for lawful exchange and fairness depend on the characteristics of the interacting families, epitomized by the exact kinship relationships involved. Given a definition of a family and its intrafamily interaction, exchange of women between families (namely, exogamy, marriage outside the family) is what holds society together. Moreover, all processes of social change and transformation are taken to be matters of such exogamous exchanges of women. Exchanges define the society, and the legal exchanges and the kinship structures fully define the social role of each person in the society.

This mode of explanation is characterized by several features, features that we shall see repeated in a variety of other arenas indicative of the power of the kinship analogy: First, whatever is not forbidden will happen, and it will happen "fully." We shall call this, following the historian of ideas A. O. Lovejoy in *The Great Chain of Being* (1936), the principle of *plenitude:* "[T]here is a fullness of the realization of conceptual possibility in actuality."[6] Second, each object is completely known by its properties, and objects are different only if they have different properties. And we know of those properties by the behavior of those objects in their interactions with other objects (namely, properties are relative). We might call all of this a principle of the nondegeneracy of names and the fetishism of properties – *nondegeneracy,* for short. As a consequence of plenitude and nondegeneracy, once we classify objects by their properties, we then may describe how they will behave; namely, how they will interact – what might be called their dynamics. For whatever is allowed will happen (plenitude), and allowedness is given by rules expressed in terms of properties (nondegeneracy). We might say: *Dynamics and classification are the same.* Finally, this kind of explanation is local and noncausal. It is local in the sense that each exchange has to be allowed and fair in itself. It is noncausal in the sense that whatever happens happens because it is possible or allowed, not because it is triggered by a predecessor event.

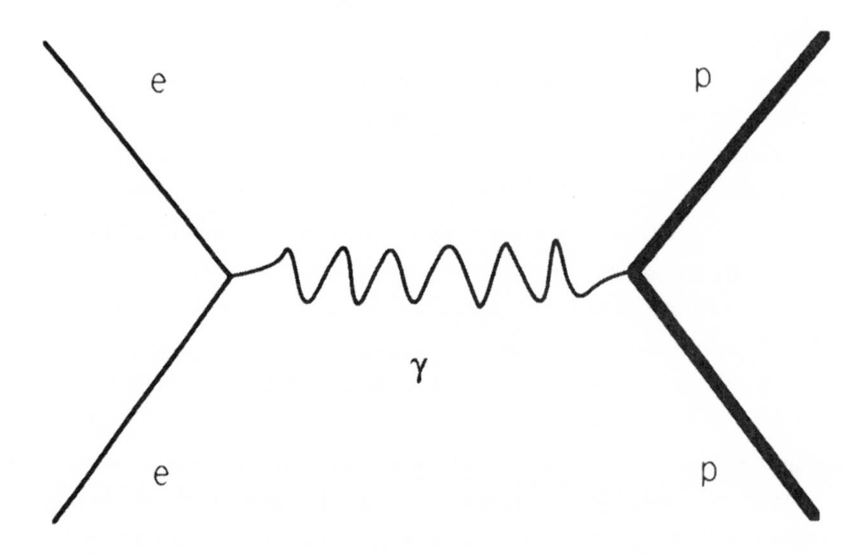

3.1. A Feynman diagram of particles interacting, e + e → p + p, an electron scattering off a proton

To the contemporary chemist or physicist, the kinship story should sound familiar, but now it is about chemical species or elementary particles in their interactions. The families are now groups or columns on the periodic table of the elements, the elements (as members of groups) chemically interacting with each other. Or, the families are multiplets of atomic energy levels or groups of elementary particles – actually in each case an extended hierarchical family consisting of smaller nuclear families – again interacting with each other. What is exchanged between families are particles such as electrons or pions or gluons or photons or W's, rather than women or money, thus allowing for the interaction (or "marriage") of individual particles. And an exchange is fair if it balances or conserves energy, momentum, charge . . . , whichever are the relevant currencies or quantum numbers. An exchange results not in marriage, but in an interaction: a reaction of chemical species ($H_2 + 2Cl \rightarrow 2HCl$ is an exchange of valence electrons as well as energy); or a collision of particles (electron-proton scattering, $e^- + p \rightarrow e^- + p$, conventionally pictured as shown in Figure 3.1, here an exchange of photons, γ's). And in direct analogy to kinship or marriage rules and incest prohibitions, again there are allowed and forbidden processes. Whatever is allowed is fully allowed. But if a charge or energy or whatever is not conserved, yet it ought to be, then the interaction will not happen, or it will happen quite rarely (something we shall discuss shortly).[7] An object's properties are its degrees of freedom; for they say how we might get a handle onto it: how it will interact, whether or not it will interact or marry. Getting the properties or degrees of freedom right means we may understand both how Nature is structured in terms of families and what Nature will allow in terms of interaction.

SYSTEMATICS IN THE FIELD

Observing a society's behavior, the fieldwork anthropologist might try to figure out kinship categories and incest taboos in order to account for observed marriage patterns (although as a start, the anthropologist could employ native informants for the everyday names and rules). Similarly, in actual practice, starting off with some everyday well-known properties and rules, physicists try to find both a set of detailed properties (masses, charges, spins, energies, etc.) to attribute to potentially interacting objects and a set of rules of exchange or interaction; so that by checking the properties and the rules, physicists can say which processes will occur, and which will not occur. Again, if the rules concerning exchange of properties are fulfilled in a proposed pairing, so an interaction or marriage is not forbidden, then that interaction will happen, and it will happen as often as we might imagine it could happen.

What is most remarkable is that those properties we impute to objects in order to account for their known fair exchanges or interactions then turn out to be versatile and robust, useful in many contexts which do not seem to be much like the original one. And so scientists take chemical valence (the number of free electrons on an atom) or the quark content of an elementary particle (originally designed to account for particular regularities of interaction but which turned out to be much more widely useful and significant) as not only imputed properties, but as real and intrinsic to the object. Eventually, we even learn to see those valence electrons or quarks, both in actual experiments and in visual models.

Technically, these imputed properties taken as real and intrinsic are fetishes. Aspects of an object surrogate for the whole, and distributed features or powers of interaction are condensed onto individual objects. Of course, from the practicing physicist's point of view, this fetishism is just what physics is. To speak of authentic wholes or of alienated objects is to do something other than physics.[8]

Physicists find ways of specifically taking hold of an object through such properties (taken as handles), or ways of interacting through exchanges of objects having such properties. Moreover, those rules of exchange also turn out to be generic and curiously prescriptive. For example, as is seen in the history of the law of the conservation of energy, the rule gives the format for such a conservation and then it encourages the physicist to find energy of various sorts so that total energy is conserved in an ever-widening range of circumstances, thereby keeping the natural world in accord with the rule.

Physicists may casually talk about families and relatives when they are concerned with problems of classification and dynamics. As we shall see, such talk is in fact fully instantiated in their modes of getting hold of Nature and setting Nature to work. They believe the world "must" go this family way (and the colloquial reference to fructiferousness is appropriate); the metaphor is authoritative; analogy is destiny – only, of course, when family and marriage are suitably interpreted. Put differently, the anthropologist's abstraction of family into kinship structure is in actual fact the powerful notion here, applying to many fields, and not the novelist's or the sociologist's abstraction of the family.

More generally, we understand physicists' obsessions and machinations once we understand what they are up to. If physicists find themselves in a family way, so to speak, they will be motivated to find the right names for relatives and the right rules of marriage.

THE PROBLEM OF "QUITE RARELY" IN A PLENITUDINOUS WORLD: SOME INVISIBLE HANDS

Let us consider one such seeming machination. It is perhaps inconsistent to say that whatever is not forbidden will happen as frequently as it might and then to speak of some interactions as happening "quite rarely" (rather than always or never). Yet some interactions do occur comparatively rarely. More generally, such intermediate or mixed cases are pollutions;

for if they are not accounted for, they defile the system of classification. How is the kinship model to be modified so that plenitude and nondegeneracy are still definitive, yet we may speak of "quite rarely"?

Immediately, we realize that if there are too few marriage partners available, then it may be quite difficult to find a free partner or to make a match. Or, if the level of wealth available is too low (corresponding to the temperature or mean energy), there won't be enough dowry available to seal too many a marriage.[9]

More sophisticatedly, one might have a hierarchy of rules about marriage or interaction which are hierarchized in strength as well. Close relatives are more forbidden as marriage partners than are distant relatives; some groups of chemical or physical interactions are distinctively stronger than are others. And often that hierarchization is exactly what is done to account for quite-rare interactions, whether we are talking of kinship or of particle physics. There are degrees of cousinhood and of incest taboo. And there is a hierarchy of different rules of interaction of elementary particles associated with a hierarchy of very different strengths of interaction – the gravitational, weak, electromagnetic, and strong interactions or couplings. In practical everyday phenomenological physics, the differences of rules and of strengths are used to give an account of what happens quite rarely.

Still, at least some of the time, some physicists are uncomfortable with this sort of artifice or device as a way of dealing with "quite rarely." They would prefer a unified set of rules and properties with no ad hoc hierarchization schemes, no matter how empirically and phenomenologically useful they are. If there are to be qualitative changes in the rules and their corresponding strengths of interaction, the changes ought to have a more smooth or continuous or unified origin. For instance, particle for particle, electrical forces would seem to be very different in behavior and much stronger than magnetic forces.[10] (For example, the electrical force between the electron and the proton in a hydrogen atom is much larger than is the magnetic force.) Yet we know that electrical and magnetic forces become more and more alike as the velocity or energy of the interacting particles increases. A more unified theory, such as Maxwellian electromagnetic theory, might show how the rules for different interactions (electrical and magnetic, for example) may be seen as belonging to just one set of rules, and so they are "the same." And a good theory will also show how the hierarchy of forces, the relative strength of interaction or coupling, changes with energy, becoming more equal at higher energy. (This might remind one of a theory that would unify liquid water, ice, and vapor, as one substance that happens to split into different phases below a certain critical temperature and pressure.)[11] And in unified theories of elementary particles, the strong (or nuclear) force becomes smaller as the interaction energy increases, and the electromagnetic and weak forces become larger – and eventually, at a high enough energy of interaction, they are all roughly the same strength and exhibit roughly the same behavior.

There is another strategy for dealing with rarity, or perhaps even overabundance, a strategy that preserves the criterion of plenitude, that whatever can happen will happen and it will happen as frequently as we could reasonably expect. Physicists say that *what we see* is a sum of a set of more fundamental processes, each fundamental or elementary process happening according to the principles of plenitude and nondegeneracy. (That is what makes it fundamental.) And in that addition there may be cancellations that give sums that are quite small such as $+3.0 + -2.9 \approx 0$, or two vectors in roughly opposite directions, $\rightarrow + \leftarrow = 0$, and so account for the rarity of some observed processes. Conversely, there may be anticancellation, the vectors pointing in the same direction, and then we have overabundance.[12]

Say we have several ways of going from A to B, yet we cannot or do not know which way was followed – in effect a degeneracy. (See Figure 3.2.) Plenitude and the demand for nondegeneracy suggest to physicists that every indistinguishable way was followed, and the process of going from A to B is to be understood as the combined effect of all the possible ways. (The different paths no longer need to be distinguished, so degeneracy is no longer an issue.) The physicist Richard Feynman describes the "democracy" of all histories or fundamental processes that contribute to what we see, each history being equally possible.[13] The combination rule turns out to be curiously straightforward. We can assign a number (called an amplitude) to each way, that amplitude representing its strength or its probability (say by squaring its amplitude), then just add up all those amplitude numbers to get the amplitude number as-

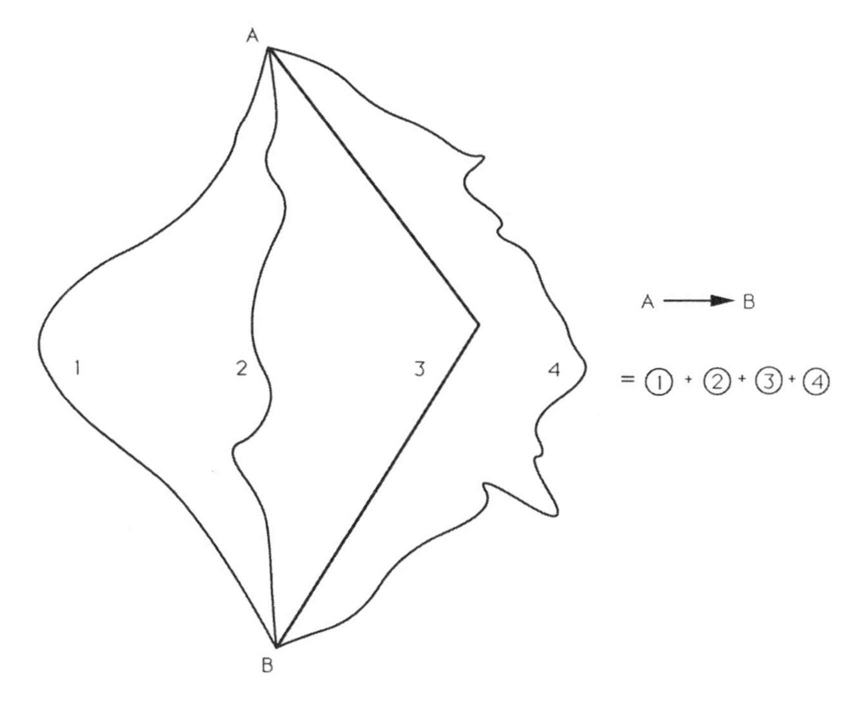

3.2. Multiple paths in a Feynman analysis of particle interactions

signed to the process of going from A to B. Again, because amplitudes may be positive or negative, or opposed in direction as vectors, some of the time that process of addition will result in cancellation, the sum being smaller in absolute value than the number for any single path. Such cancellation is called interference, in honor of the prototypical case of the interference of light from two sources, in which the pattern of dark and bright shows the consequences of such cancellations and anti-cancellations. Just because we allow a signed number or a vector to represent each contributing indistinguishable path or process, there can be summation effects indicating that what we actually observe is quite rare.

Physicists employ the idea that rarity or overabundance is a secondary consequence of essentially plenitudinous fundamental processes in quite subtle and inventive ways. Consider a gas of atoms or molecules in equilibrium.[14] Technically, it turns out that quite pronounced cancellation effects among all the possible microscopic states or arrangements of the atoms are the mechanism for identifying a crucial feature of the unrare equilibrium state, the existence of a temperature, T.[15] (Here, equilibrium is *thermal* equilibrium.) Once they have so identified the temperature, physicists again retain the principle of plenitude, now saying that at equilibrium all possible atomic or molecular arrangements will occur, but with a probability or weight factor (much like the amplitude number above) that depends on the arrangement's total energy, E, and the ambient temperature, T: $e^{-E/kT}$ where k is Boltzmann's constant, and $kT/2$ is called the thermal energy, the mean energy of a particle, actually of a degree of freedom, at this temperature.[16] (This is a "quite rarely" if the energy, E, is much greater than kT; the factor $e^{-E/kT}$ is then close to zero.) Now, rather than plenitude being justified as a democracy of histories, it is justified by an ("ergodic") hypothesis that all possible arrangements must appear.[17] The exact details of what we actually see, such as pressure or specific heat, are said to be measured by a sum of the weights of all the possibilities (rather than a sum of the amplitude numbers, as in the democracy of histories), or functions of that sum. This sum is called the "partition function" – namely, a function of all the possible ways of partitioning a system, all the arrangements of its parts. And it turns out that that sum is usually dominated by arrangements whose energy is about $kT/2$ per degree of freedom.[18] (Note that we have invoked plenitude twice in this paragraph: first in identifying T, then in forming the partition function.)

Moreover, since there are many microscopically different arrangements of the component atoms, many if not most of which look the same in the everyday macroscopic world, we would seem to be haunted by degeneracy. But, again, physicists say that what we actually see, the actual equilibrium of the gas of atoms, is the one macroscopic situation that is instantiated by the *most* possible different, yet macroscopically indistinguishable, microscopic atomic arrangements. Degeneracy becomes both a virtue and a non-problem.

In each case, by quite subtle device, seeming nonplenitudinous effects are taken as secondary consequences of the application of plenitude at what is taken to be a more fundamental level. Yet these devices are not just jerry-built tricks; they are finely wrought and rational machinery.

They turn out to be foundations for quantum mechanics and statistical physics. More generally, in modern society we are entranced by an ironic theme: that the mechanism through which the world actually works and what we see phenomenally are in poignant contrast. The gears move wondrously and what we hear is a simple tick-tock. As in the Invisible Hand of market processes and in Hegel's account of historical processes (the Cunning of Reason), the world works in backhanded yet ultimately orderly ways. Here, there need not be a formal identity of the microcosm and the macrocosm.[19] So physicists have come to expect that a system may not be like any of its parts (although as we have seen in the previous chapter they may still aim for such a homology). Rather, there are to be surprises that are the consequences of plenitudinous cancellation and degeneracy, namely, unexpected effects of combination.

MARKETS AND FETISHES

Adopting the model of family and kinship, our commitments are: (1) to processes of exchange (there has to be a match at either locus in the exchange process, so in physicists' description of scattering, diagrammed as shown in Figure 3.1, conservation laws are obeyed at both the left and the right vertices, or in an economy both buyers and sellers have to be satisfied); (2) to rules of allowedness based only on the properties or degrees of freedom of the exchanged objects and the properties of their origins and destinations, those vertices; and (3) to a principle of plenitude, that whatever can happen will happen, fully. We then have to find the right processes, the right rules and the right properties or nondegenerate names, and the right way of understanding that everything that is possible will happen – "right" in the sense that we have an accounting of how Nature goes about its work, one that is in accord with our sophisticated yet still intuitive notions of what family as kinship structure is to mean; and one that is in accord with how we have used these models before, traditionally and conventionally, as in the periodic table of the elements and chemical valence.

Nowadays, as I have already indicated, the most palpable mechanism possessing these features is perhaps neither kinship nor particle physics. Rather, it is in market economies that most of us are aware of how processes of exchange, consumer satisfaction or sovereignty, a fetishism of commodities (treating them as fully characterized by their individual properties), and pervasive information and freedom of action – the features outlined in the previous paragraph, and were they to be the case in fact – lead to a world that is taken as stable, and organized, and efficient or optimal. When a market is in equilibrium, there exists a set of properties (namely, prices) that lead to exchanges that, arguably, are best for everyone.[20]

The political problem in early modern society was to set up social relationships so that there is a kind of exchange and freedom that allows for the operations of a market.[21] The task inherited by those who wish to use market explanations for social activities such as marriage or law is to define the appropriate exchanges, prices and properties, and modes of interaction, so that a market mechanism in its operation describes what people actually do.

Now, given the association with market economies, it is perhaps not surprising that the family and kinship mode of explanation is an epitome of the fetishism Karl Marx decried. For if things are fully known by their properties, properties taken as intrinsic to them yet which we learn about through exchange, such things may then be divorced (or "alienated") from their social context and meaning. Now in actual life, we know other persons in much more complex and rich ways than only through exchanges. But, for elementary particles, we know them only through their exchanges or interactions and hence their properties. And that is just how physicists prescribe they are to be known and to exist. The actual specific kind of particle properties is determined by the classification system that is chosen or, in a different vein, the division of labor.[22] There could be alternative assignments of kinds of properties. But once a division is made, and it works to produce Nature, we then forget the designing or theoretical work we have done; and we take particles as fetishes, as just having these properties. Of course we may then discover – empirically, doing an experiment! – that some properties are distributed elsewhere, yet we think they should be attached to certain particles. For example, the neutrino's behavior turned out to depend on whether its sibling was an electron, a muon, or a tau particle. In response, physicists attached a new property to each neutrino – the name of its sibling: ν_e, ν_μ, ν_τ, eventually creating new family structures – and so, by the way, insisting on nondegeneracy and fetishizing the world once more.

In sum, whether it be kinship, particles, or markets, the family and kinship model encourages us to put our faith in the interaction of a set of individuals, individuals defined by their role in a family, families defined by their role in nets of extended families and in society, each interaction prescribed by the interacting individuals' and families' properties, and each interaction being an exchange process. Nature is not any single interaction, but the composite of all allowable interactions. And each individual is fully known through those properties it possesses (namely, that we may attribute to it). Here, freedom is identified with necessity through the use of "all" and "fully." Whatever is possible must happen; whatever some thing can do is prescribed by how it is to be recognized and known. The individualism and freedom characteristic of classical liberalism lead to or follow from an account of how the natural world works and is organized.

TAKING THE RULES SERIOUSLY: SURPRISE AND EMERGENCE

Again, our experience is that Nature often works in a surprising fashion through freedom and necessity. These workings and the effects of combination, as phenomena, really are surprises, were we only given a set of rules and properties and taboos. Neither Nature nor markets are obvious consequences of exchange as the means of interaction of their parts. Now some of those phenomena were of course known before we had an inkling of the rules and mechanisms which could be used to generate them. And rules are designed by physicists to give an account of a subset of those already known phenomena. Still – and this is the real surprise – there are often predictions of new phenomena that are then validated, and explanations of known phenomena which were not part of the original subset. The characteristic attitude here is the physicist's willingness to take seriously a set of rules and properties – to be committed to exchange and fetishism, for example – and see where it leads.

Taking the rules seriously, the world must go this way of plenitude, nondegeneracy, and exchange – although, of course, there is room for ingenuity in interpreting the rules, in figuring out the right names or degrees of freedom, and in surmising those ironic effects of composition.[23] And there is also room to empirically find phenomena not at all obviously encompassed by the rules. The physicist is here working within a sufficiently constrained system so that invention and discovery become interesting.[24]

Quantum mechanics is a story of freedom and necessity and, as we put it earlier, democracy. Again, the governing rule is that everything that can happen will happen, unless it is specifically forbidden. And, again, there is a principle of nondegeneracy: If we do not see a difference, there is no difference; and if purportedly different objects have identical properties or handles or labels they are the same. Moreover, to preserve the principle of nondegeneracy, if we do not actually take hold of or observe some properties, properties which we otherwise know about from other attempts at handling or exchange, then all the various possible unexamined properties or labelings or histories are to be added up – that is, their amplitudes are to be added up in figuring out what we do observe.[25] Put differently, "not looking" might well be empirically interesting. Recall that in such an adding-up, one might get interference effects, and these may well be of some consequence. Taking the rules of plenitude and nondegeneracy absolutely seriously, physicists find they are rewarded with surprises and with opportunities for invention.[26]

The story of the *K*-zero (K^o) particle system is one such providential reward, offered by physicists as testimony to the truth of quantum mechanics and its commitment to kinship rules. Most often produced in particle accelerators by the collision of the beam with a target, the *K*-zero is a particle with no electric charge (a neutral particle) that is about 1000 times as heavy as an electron and half as heavy as a proton. It radioactively decays in a variety of ways in about a billionth of a second. (As we shall see, it exhibits two lifetimes, the second being about 600 times longer than the first.)

Now in telling this story of the *K*-zero, my purpose here and throughout the book is not so much to instruct the reader in detailed physics or in the workings of Nature, but rather to present the rhetorical moves

$$K^0 \xrightarrow[a]{} \pi^+\pi^- \qquad \bar{K}^0 \xrightarrow[a]{} \pi^+\pi^-$$

$$K_1^0 = (K^0 + \bar{K}^0)/\sqrt{2} \qquad K_1^0 \longrightarrow \pi^+\pi^-$$

$$K_2^0 = (K^0 - \bar{K}^0)/\sqrt{2} \qquad K_2^0 \not\longrightarrow \pi^+\pi^-$$

3.3. The K^o system

physicists make when taking hold of the world. What follows is mildly technical. I have also been schematic and ahistorical, leaving out some wonderful details about the motivations for the right degrees of freedom.[27] What should be noted are the lines of argument and the kinds of moves that are thought to be satisfying in the family and kinship analogy.

About half of *K*-zero particles are observed to decay into two pion particles (say, $\pi^+ \pi^-$; see Figure 3.3), each of which is about 280 times as heavy as an electron, with let us say an amplitude number *a* for that path or history; and about half of *K*-zero-bar ($\bar{K}^o$) particles, the antiparticle of the *K*-zero, also decay into two pions, with the same amplitude *a*.[28] Now if we were to see just those two pions, as a radioactive decay whose total energy is that of the mass of a *K*-zero (or *K*-zero-bar – the masses are the same), we would not know if their parent was a *K*-zero or a *K*-zero-bar. When we do not know which was the parent particle, we might well sum both histories (weighting each equally) to find the amplitude for two pion decay. Moreover, nondegeneracy suggests (and this suggestion actually required a lovely piece of invention to be realized) that we define another particle, called the *K*-zero-one (K^o_1), as a weighted sum of the *K*-zero and the *K*-zero-bar, and the *K*-zero-two (K^o_2) the difference – where by the sum and difference of particles physicists practically mean the sum and difference of the particles' amplitudes for a particular history. Now we have one particle, the *K*-zero-one, that decays into two

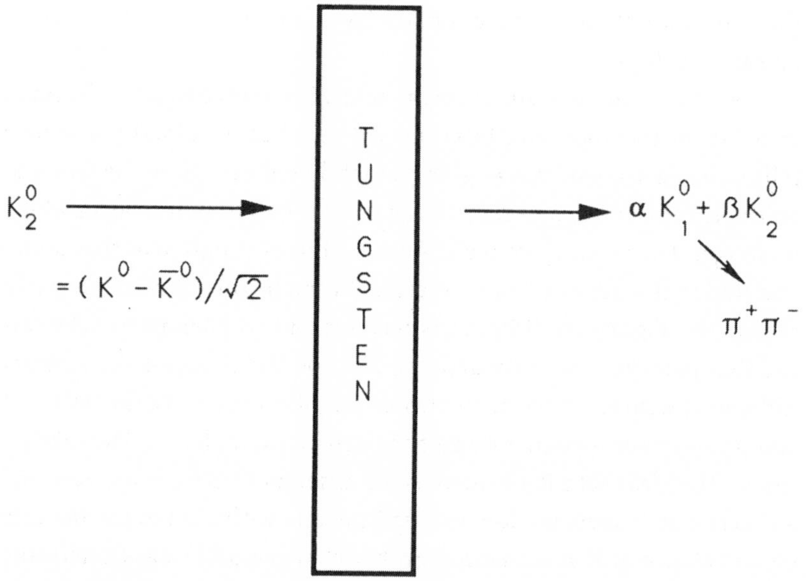

3.4. Scattering of K^o particles in tungsten

pions ($a + a$ is not in general equal to zero), and another, the *K*-zero-two, that does not decay into two pions ($a - a$ does equal zero). (In effect, the amplitude for two-pion decay really belongs to the K-zero-one, and not to the K-zero or the K-zero-bar.) The *K*-zero-one has an easy marriage partner or decay path, one that is unavailable to the *K*-zero-two. From our discussion of "quite rarely" we might expect the *K*-zero-one to decay more rapidly than its sibling. And in fact two lifetimes are observed for *K*-zero particles.

Now we have a way of accounting for why only half the *K*-zeros decay into two pions (a *K*-zero is only ½ *K*-zero-one). Moreover, unlike the *K*-zero and *K*-zero-bar, here, for *K*-zero-one and -two, different things are done by different particles, and whatever each particle does it does all the way. (The K-zero-two decays into three pions.) So far *all* we seem to have done is some relabeling (hopefully using the right degrees of freedom, so we won't be sorry) following the rules of nondegeneracy and plenitude.[29] We have put what we know in a more acceptable form. And as our empirical payoff, we account for why the *K*-zero has two lifetimes:

there are actually two different particles in a *K*-zero beam, the K-zero-one and the K-zero-two.

What is even more surprising is that it is possible to take a beam of pure *K*-zero-two (obtained by taking a *K*-zero beam and waiting about a billionth of a second, that is, going a foot downbeam from the target to let the *K*-zero-one component decay away) – this is, now a beam which no longer decays into two pions – place a foil of tungsten at that point, and find that when the beam emerges it then begins to decay into two pions. (See Figure 3.4.) Physicists account for this phenomenon by saying that particles and antiparticles, *K*-zeros and *K*-zero-bars, interact differently with ordinary matter (as we might expect, since the foil is not composed of antimatter, not anti-tungsten for example). And we take seriously the idea that a *K*-zero-two is the difference between a *K*-zero and a *K*-zero-bar. So when a *K*-zero-two interacts with matter, the delicate equal balance of *K*-zero and *K*-zero-bar changes, and so arithmetically and actually some *K*-zero-one is now present (it is "regenerated") – and it decays into two pions.

Any sensible reader might find the previous paragraph beyond belief, or at least a fudge. Yet, in fact, the analysis does work, in great detail; and it is a prototype of the kinds of explanation quantum mechanics offers and of the phenomena it accounts for. What is called for is a commitment to plenitude, to nondegeneracy and to believing that some degrees of freedom are the right ones, and to the combination rule provided by the democracy of histories (which is an instantiation of plenitude and nondegeneracy). Cancellation of amplitude numbers really means that some things won't happen; and relabeling to remove degeneracy has real consequences. Names matter. Taking plenitude and nondegeneracy seriously, and so not shying away from the atavistic quality of taboo and incest and family, the various so-called paradoxes of quantum mechanics turn out to be just surprises.[30]

Another of the surprise-filled stories of freedom and necessity are those of evolution. Evolutionary mechanisms describe the development of the system of organisms in terms of the interaction of entities (genes, organisms, families, populations, and species) – in the end, literally an exchange of genetic material, sometimes mutated, within varied environments. Just about any exchange that is conceivable will happen, conceivable in the sense of species, but only those interactions that produce more viable entities will have persisting effects (just which type of entities being a matter for some argument). Eventually, organisms with qualitatively new features appear, their qualitative novelty apparent in the context of other similar creatures. Such "emergent" phenomena depend on our seeing all organisms as part of a coherent system as in Linnaean taxonomy, so that surprise, which depends on a comparison of naive expectation and actuality, is possible.[31]

Also, we may be surprised by the effects of interaction if our expectations are that organisms in a system will differ from each other incrementally if the process of exchange is incremental. In this family and exchange model, classification is expected to be the same as dynamics, a dynamics which would seem to be incremental. Yet in biological systematics or taxonomy, classification in general depends on comparatively sharp differences. But since the system of exchange and interaction occurs at various levels of organization and complexity (genes to species) and fluctuations are influential, and since environmental influences on an organism or collection of organisms, such as geography and topography, affect genetic composition over the long run, incremental exchanges and dynamics may well lead to global differences that are comparatively discrete and discontinuous, and hence we might have speciation.

No matter what the arena of concern – kinship, particle physics, economy, quantum mechanics, or evolution – there will always be surprises.[32] Those surprises are, as I have indicated, a consequence of our starting out with elementary parts whose possibilities for making up a complex system, say through hierarchization or coherent addition, are much more varied than we might have expected given both the simplicity of the parts and rules and the seeming innocence of nondegeneracy and plenitude as principles. The structure of the complex system need not be at all homologous, at least when we first look, with the structure of the system of labels and rules and objects.[33]

Again, it should be noted that the phenomena to be explained have something of a life of their own. Markets and chemical reactions came well before their detailed explanations, although perhaps some of their crucial features were first discovered and described using the theoretical structures that go along with plenitude and nondegeneracy and exchange. The phenomena are surprising in relationship to the meager parts we use to understand them. But those meager parts were invented to do some of that surprising work.

STRUCTURE AND SYSTEM

For physicists in the grips of the family and kinship model, it actually is gospel that you shall be known (fully!) by your deeds and fruits, where those acts evidence plenitude, nondegeneracy, fetishism, and exchange. More specifically, the principles behind freedom and necessity are:

– Everything that can happen, will happen. And it will happen with the same frequency (or the same probability) as anything else that can happen. Or, to repeat Lovejoy on *plenitude,* there is a "fullness of the realization of conceptual possibility in actuality."

And what we could conceive of but does not happen, is actively forbidden.

– By the principle of *nondegeneracy,* if two things act differently in the same circumstances they are different; and if they have different properties, there must be at least some circumstance in which they do act differently, a way of getting a handle onto their difference. Properties must make a difference if they are to be real properties.

And so a name fully describes what something is and what that thing will do in relationship to other things (*fetishism*). (For more on names see chapter 1.)

– Nature is to be produced by the *system of interactions* of elementary parts or individuals through all the allowed processes of exchange, "elementary" meaning that the parts' properties are nondegenerate. Each interaction *conserves* or balances the properties of what is exchanged, whether they be energy or market value, for example.

Nature is interaction.

– Rules for interaction are simple. *Complex Nature is a product of simple rules.* Rules (and parts) are adequate to their task if, when they are applied everywhere, what they say is possible is just what we see, and what they say is forbidden is not seen. If the rules are not simple, or the names are degenerate, or the rules prove inadequate, it is taken that the elementary constituents are not well understood.

–If there is plenitude, nondegeneracy, fetishism, and a system of interaction, then it is natural to believe that structure and system are the same, that *classification and dynamics are the same*. Names are used both to classify parts into families and to specify rules for interfamily interaction. Dynamics indicates classification; classification prescribes dynamics.

These are remarkable principles. But they are the way the world must go once we take it in terms of family, kinship, system, and exchange. These commitments pervasively inform what physicists do and believe. More generally, as commitments they are often accompanied by ideological justification and philosophical explanation of why there must be exchange, properties, and systems, often a story of individualism leading to solidarity, or of fetishism and equilibrium, or of economic comparative advantage and brotherhood.[34] They are taken to be generic principles, and a specific candidate for the truth must be in accord with their form and requirements.

The principles are, as well, a story of freedom and necessity. Here freedom is the chance to fulfill any possibility that is in accord with the rules. Necessity is the fact that in the end there must be only one overall possibility: *this* world. (More generally, this is a story of revelation, the particularity of Providence.)

In a manner at first seemingly different from what Adam Smith describes, a collection of motley individuals becomes an articulated culture – or an economy, or the physical or chemical world – when we invoke the various rules of forbiddenness. Physicists speak of "turning on" those rules and taboos, since they conceive of what it might be like were those rules absent, a fundamental abstraction.[35] Turning on the rules and taboos now allows the individuals' putative properties to distinguish individuals from each other, hence classification, the divisions of labor, and the story of parts. Put differently, law creates actual society, or as Freud put it, we have a story of civilization and its discontents. In such a culture or economy or Nature, individuals are to be properly labeled so that they can follow the rules, and in following the rules they are free.[36] And in so being free they work out the possibilities of the society or system.

This set of prescriptions informs modern economy, politics, culture, and science; and it sometimes seems there is nothing else, or so we might let ourselves believe. Since there really is something else, including violations of taboo, unfair exchanges, disequilibrium, stickinesses, and leftovers (as in the exchange of gifts and in the web of our mutual obligations) – fluctuations, to speak as a physicist might – the actual story has to be somewhat richer.[37] How we might begin to give such an account of the Factory of Nature, and still insist that Nature must go this way, is the story of our next chapter.

FOUR

The Vacuum and The Creation: Setting a Stage

So Far, an Epitome; Sweeping Up the Vacuum; Symmetry and Order. The Empty Stage; Of Nothing, Something, and the Vacuum. Setting Up the Stage; Ideologies for a Vacuum; The Dialectic of Finding a Good Vacuum; The Analogy of Substance, Once More. Fluctuations in a Vacuum. Annealing the World.

THE ARGUMENT IS: THIS CHAPTER DESCRIBES A SET OF STRATEGIES employed by physicists for achieving a comparatively simple world: They find an orderly emptiness – a "vacuum" that is much like our everyday notions of an empty space, properly understood. And in that vacuum, degrees of freedom that are in accord with that orderliness show themselves most effectively. Other degrees of freedom are tamed or repressed, and so they are hidden, and it is just in this sense that the space is empty. That is how the vacuum can both be empty and exhibit simple, orderly phenomena. (For physicists, a model for this vacuum is an orderly crystal.) This strategy might be called *theatrical,* a matter of setting a stage on which only certain actors and actions may appear. Along the way, an account is given of how Something that is in accord with this orderliness arises out of Nothing, an account of creation.* And that arising is an abrupt or discontinuous or fairly sharp transformation, much like freezing.

*Something and Nothing (capitalized) are used here in a very particular sense, as what populates a vacuum and what there was "before" it existed (when things were hotter or earlier or more symmetric), respectively. But I also mean to provide the physicist's answer to the rather more generic question of how something arises out of nothing.

The strategies I describe here and in the next chapter are rhetorical and analogical, and physical and ontological: for, whether it be Nature or Scripture, it turns out that we face the same problem in describing how an orderly world of something arises out of a soup of disorder and chaos and nothingness. And so those who describe the mundane and the sacred worlds use similar rhetorics and borrow analogies from each other. That the world must go this way would seem to derive as much from the grammar of "nothing" and "something," as out of Nature and physics and mechanism. And the obsession with creating an orderly vacuum or an orderly social world is a generic one about origins and creation. For if you get the Big Bang or Genesis right, the structure of the subsequent physical and historical world will then make sense.

Now to have mastered the various rules and strategies, and their associated obsessions and imperatives, as I have been describing them in these chapters, is not enough to make you a physicist – although if you are a physicist you will share those obsessions, you will share in that subculture's way things must be. Rather, you become a physicist by your actual involvement with the *particular* problems and material physicists take as interesting, and along the way you pick up the rules of play. The obsessions and rules are rarely argued for explicitly. As I have said, each is taken as an "of course," and as such, they collectively reveal the foundations of the enterprise. To do physics, one participates in its practices, culture, and ideology, thus employing the conventional models and analogies.

I

SO FAR, AN EPITOME

We say: The world must go one of these ways, one of the ways it has gone before. So we (as physicists) believe that doing physics is a matter of getting the division of labor right and a matter of understanding how the factory or the economy works. And we find walls and particles and fields that together produce Nature. Or, we believe that the natural world and the factory and the economy are mechanisms, ones that can in principle be taken apart and put together again, and again our problem is finding the right parts and figuring out how they work together. The mechanism may be like a random walk, or a spreadsheet, or the calculus, or a bureaucracy, or a computer program – or even a clockworks. Or, the world is like a society of families or a complex market economy, governed by rules of marriage and incest, of allowedness and forbiddenness of exchange. And what is not forbidden is allowed; and, eventually, it will happen. And the account of classification is the same as the account of dynamics.

Corresponding to each of the ways that the world must happen are superordinate values. To the factory there is fit and efficiency; to the mechanism there is the simplicity and versatility of its parts and the robustness of its organization; and to the family and kinship there is solidarity and the fairness of exchange. And in order to make sure that the world goes these ways, and the values are fulfilled, we are committed to particular conventional forms, whether they be particles, fields, and walls, or differential equations, or structures of classification and modes of exchange.

Some notions, such as fields, particles, and interactions, do keep reappearing in these various ways of taking the world. But the notions play somewhat different roles, doing somewhat different work, in each way or model. A particle conceived of as an enclosed object with some leftover degrees of freedom might be shown, by mathematical and physical argument, to be the same as a particle conceived of as something that interacts through exchange with other particles. Yet in each context we have different expectations of a particle, of what it must do and how it will do that work, although again those expectations may be intimately if subtly connected. And it is those expectations, at least as much as the abstract notion of a particle, that guide us in its employment. Of course, physicists may shift approaches and so their expectations, at one moment thinking of a particle as something that holds in many degrees of freedom, and at another as its being an object whose properties determine its interactions.[1] Physicists do not seem to think so much abstractly about their tools or notions as they think of them in the specific contexts of their employment – just as we usually think of words in the context of phrases rather than in the context of a dictionary. Those differences among the contexts matter enormously. To be an effective factory designer, craftsman, or classifier, you have got to know how to use each notion, just right, *in situ*.

In these next two chapters there will be two more stories of the ways the world must go, with their consequent obsessions and commitments. Here, in chapter 4, the world is conceived of as an almost empty stage, with only a few actors upon it, with only a few degrees of freedom active – the rest of the world, the rest of the degrees of freedom, being tamed or being hidden behind the sets. Of course, we know full well that there is lots there, that the emptiness hides a great deal. But the emptiness of the set stage really does effectively hide what is backstage. Put differently, in chapter 4 we shall be concerned with how to design both a factory and its outside suppliers so that the factory's production process is straightforward. In chapter 5, the world is taken as something to be poked at and handled, the knowledge we have of it being a matter of how we do that poking and handling. The image there will be of tool and craft. And the obsession will be to arrange things so that when we poke at something, what we see in response is simple and distinctive. As industrial engineers, we will want to analyze the factory's workings by employing sensitive probes.

SWEEPING UP THE VACUUM

Let us now set out the task for this chapter, which in many ways might be taken as commentary on Abraham Pais's description of Einstein: "Better than anyone before or after him, he knew how to invent invariance [or symmetry] principles and make use of statistical fluctuations."[2]

As we have often said, in their conceptualizations and in their experiments physicists hide many of Nature's degrees of freedom.[3] Those hidden degrees of freedom are made to be, or have been shown to be, irrelevant to a situation just because of its mode of being set up. The details of particular molecules' positions and momenta are irrelevant to the properties of a gas in thermal equilibrium. The particular molecules' arrangements in a magnetic material are irrelevant to its properties when, at its critical temperature, it becomes cool enough to be permanently magnetizable. (In fact, at the microscopic scale and larger, the molecular arrangement then looks the same at all scales – it exhibits a "scaling symmetry.") And many of the elementary particles to be found when we look within and so vigorously shake up the nucleus are hidden, frozen out of consideration, in low-energy nuclear phenomena, close to the lowest energy or "ground state." These situations, defined by equilibrium, scaling behavior, and the ground state, respectively, hide lots of degrees of freedom, a hiding that makes for an absence or an emptiness.[4] By such hiding, physicists define a simple and orderly world. They define what we shall come to understand as a structured emptiness – a *vacuum* – which has a basic orderliness that then delimits both the objects that can populate that vacuum and those that are banned (and hence the vacuum is empty, absent of the latter).

More generally, the prescriptions for vacuum creation that physicists employ are ways of taming many degrees of freedom. As we shall see, in their making of a vacuum, physicists set a *stage* on which action makes sense, lines of motive and causation are clear, origins are accounted for, and the filiations of structure or organization are readily discerned. The physicist's motive might be taken to be "sweeping up the vacuum," cleaning out more of the leftover parts:

> I realized that I had to think about the degrees of freedom that make up a field theory. . . . to eliminate an energy scale or a length scale or whatever from a problem [so as to thin out the degrees of freedom], . . .
>
> – K. G. WILSON[5]

> [You want] a way to arrange in various theories that the degrees of freedom that you're talking about are the relevant degrees of freedom for the problem at hand. . . . [Y]ou are arranging the theory in such a way that only the right degrees of freedom, the ones that are really relevant to you, are appearing in your equations.
>
> . . . [Y]ou concentrate on the degrees of freedom that are relevant to the problem at hand. As you go to longer and longer wavelengths you integrate out the high-momentum degrees of freedom because they're not of interest to you [in studying phase transitions] and then you learn about correlation functions at long distances; or, vice versa, you do what Gell-Mann and Low did, and as you go to shorter and shorter wavelengths you suppress the long wavelengths. But sometimes the choice of appropriate degrees of freedom is not just a question of larger or smaller wavelengths, but a question of what kind of excitation we ought to consider. . . . [Rather than a matter of scale you're] actually introducing *new* degrees of freedom as you go along [as in the superconducting ground state, Cooper pairs].
>
> – S. WEINBERG[6]

> One invents an effective medium in which the *average* scattering due to the random objects one is looking at is exactly zero, the remaining scattering then being at least incoherent and often small: it is a clever scheme and an effective one, if what one is concerned to do is to sweep the dirt in our dirty system under the rug and think of it almost entirely as a perturbed clean one.
>
> – P. W. ANDERSON[7]

SYMMETRY AND ORDER

Warning! I should note immediately that the terms *symmetry* and *orderliness,* as they are used by physicists, reflect opposing tendencies: either toward greater ignorance and invariance, or toward greater knowledge and particularity, respectively. They are comparative and conditional terms. Comparatively, a situation is said to be more symmetric and less orderly than another. A gas is more symmetric than is a liquid or a solid, for it looks the same (it is said to be invariant) after a greater variety of transformations, such as kinds of movement and interchanges of its atoms; while a crystalline solid is the least symmetric and the most orderly, most sensitive to movement and interchanges of its atoms, for you might then change its crystalline structure and so it would look different. In general, a more symmetric system is comparatively unbiased, lacking in preference, reflecting a greater ignorance of its detailed features. Conditionally, physicists define a symmetric system as one that looks the same after a specific set of changes or transformations. Mirror symmetry means that a left-right interchange – that is just what a mirror does – leaves things the same. We could not know which side we are on. And a crystal's symmetry means that if we rotate or translate it a certain amount it will still look the same.

Orderliness implies specific knowledge, a distinction or a specific direction rather than arbitrariness. A crystal is orderly in just that repetitiveness of its structure, which in effect both points in at least one direction in space, the directions of repetition, and breaks up the homogeneity in that direction by its discrete predictable repetitiveness.

Orderliness is a consequence of the choosing of a specific value for an indicator or a pointer. As a model of such choosing, physicists imagine a table on which a pencil is standing on its end, a pencil which then happens to fall down in some direction "spontaneously", as physicists say (and hence the particular direction of fall is not predictable). They then get a very great deal out of such an image. Let us say that Nothing is that symmetrically balanced pencil, the nothingness of the world being that lack of horizontal direction. The emerging vacuum (which is *not* Nothing) is the choice of direction, so breaking the symmetry; and Something, the consequent appearance of particles or matter or radiation in accord with that orderliness, is perhaps represented by the length of the pencil's horizontal projection.[8] The vacuum is less symmetrical and more orderly than its origin, the Nothing from which it fell. (Of course, that Nothing may well be an orderly vacuum itself, with respect to another even more symmetric state. Here, Nothing and Something are always in comparative relation to each other.) In this account, we get something from nothing by the falling of a pencil, so to speak – the breaking of a symmetry. So, for example, the Big Bang itself might be presented as the breaking of a symmetry.[9] And the beginning of mundane historical time is presented in Genesis as the Fall.

Now, in everyday parlance we conflate order with symmetry because an orderly system is also symmetrical – the repetitiveness of a crystal meaning that it would look the same after we translated it one unit. Also, such comparisons are always relational: a crystal is more orderly than is a liquid; it is less orderly (and more symmetric) than is an imperfect crystal, say having one molecule out of place so it is a not-so-repeating arrangement of molecules. In general, when something is cooled down, the frozen state is more orderly and less symmetrical than the liquid. And in heating it up and melting, there is less order and greater symmetry.

II

> All the world's a stage
> And all the men and women merely players
> They have their exits and their entrances,
> And one man in his time plays many parts, . . .
>
> – SHAKESPEARE, *As You Like It*

Having concluded the preliminaries, I shall begin the chapter's description with an account of a theatrical stage, the Creation and Genesis, and the physicist's vacuum; then turn to the physicist's cosmological story of creation (the Big Bang) and the physicist's mundane story of phase transitions such as the freezing of a liquid; and, along the way recall the wall-enclosed particle – which together, as we shall see, have rhetorical, formal, and physical analogies. What will be crucial to our account is what I call a rhetoric of theater, an analogy of substance, a discontinuity of transformation, a hierarchy of order, and an ontology of liminality.

I shall be indulging in a scriptural and a theatrical rhetoric, for in the story the Hebrews crafted, in theater, and in natural science, people have to invent both what I call the Nothing (the world "before" creation) and that orderly Something after. Moreover, the physicist's account of creation still must somehow answer the conventional questions about how something arises from nothing, and about the origin of the orderliness of our world.[10]

THE EMPTY STAGE

> In the beginning God created the heavens and the earth. The earth was without form and void, and darkness was upon the face of the deep. . . . And God said, "Let there be light"; . . . and God separated the light from the darkness . . .
>
> – GENESIS 1

Everyday environments are in general too messy and complex to be stages for theater. Rather, we might want an empty set stage, a stage that is initially black and silent, and then we might enlighten it and people it. But this was exactly God's task and situation in Genesis, setting the universe in motion. Moreover, it has been argued that the setting of that stage – in the sequential divisions of darkness and light, sea and land, animals and humans, men and women, and so forth – provides just the needed classifications that then specify the possibilities for social interaction in the Israelites' society.[11] Here, the Creation *is* the creation of social structure (and so we recall the realm of the family and kinship of the previous chapter). What the Creation creates is orderliness – a structure which is then played out in historical time. And that is exactly what a physicist's vacuum turns out to be.

But, for the moment, let us briefly return to that empty theatrical stage. Whatever happens takes place on that stage, or is perhaps just offstage. And if the stage is properly designed and lit, the actual action will be natural and effective. Everything that could not be on stage, or at least not be referred to there, may be ignored; for the purposes of theater it is declared irrelevant. And everything that is on stage is highly stylized and archetypal, formal and structured.[12] In effect, we have abstracted and arranged everyday experience into a cultural artifact, what I will call a vacuum (an orderly emptiness) and Something within it. This is a very great act of abstraction and hiding; but in so doing, dramatists (and physicists and the God of Genesis) are setting up a world in which what ordinarily overwhelms us in everyday life becomes rather more comprehensible and meaningful.

In creating a simple physical world – a vacuum and its energetic excitations (Somethings or particles) – physicists in effect reenact Genesis and the Creation, and, as we shall see, the physicists' own story of creation in the Big Bang: all by designing a stage and delineating the action to be played out upon it. Rather than the factory, the clockworks, or the family, here we have theater and stage design. Designing that stage is a matter of inventing the right sort of emptiness. Moreover, much as the playwright and the director work within a tradition of plays and performances, to be a physicist is to be committed not only to theater, but to particular and conventional and traditional ways of obtaining that empty stage or vacuum, all as ways of creating an orderly physical world.

Now a vacuum, as I have described it, is much like the wall-enclosed particle, the wall hiding a very great deal. The makeup of a particle or a vacuum or the stage is quite rich and complex. There is a lot on the other side of the wall, below the floors, and behind the scenes. But to those of us who walk into this theater, it really is comparatively dark and quiet and empty. And then there is Something there, such as a well-defined particle, or particles interacting with each other or vibrations of a crystal lattice, or a good play, or this world or this universe – all being the consequences of creation.

For the physicist, creation – whether it be creation of the universe or of a vacuum – is usually analogized to some discontinuous transformation (recall the spontaneously falling pencil). But creation is not only analogized to that transformation, it is for some situations just that transformation. Rather than creation being a verbal device ("God said" or stage directions), here the device is a physical mechanism – whether it be freezing and crystallization, or becoming permanently magnetizable, or the setting in of equilibrium. The vacuum will then be the orderliness of a crystal, or of a permanent magnet, or of an ideal gas or fluid or an atom. In the case of the crystalline solid this vacuum tames the component molecules' degrees of freedom, since the molecules are now stuck roughly in their lattice positions, the lattice now vibrating as a whole.[13] And, Something is taken to be the acoustic or elastic vibrations of that crystal lattice as a solid whole. Now, which analogy to creation is the appropriate one depends on the specific physics of the situation. Each model system provides an account of a comparatively empty orderly stage created by a discontinuous break, and of its sequelae. If we choose the right transformation to represent our particular situation, we can then set up a sufficiently simple theatrical world.[14]

Again, the fundamental problem for the physicist is to find a good vacuum for a specific set of circumstances, an account of the prevailing orderliness of the world. Once the physicist has found such a vacuum, "irrelevant" degrees of freedom are automatically hidden, by the way of their being part of the emptiness – and the physicist (and Nature) can go to work. That vacuum's degrees of freedom are seen in the Somethings (the matter, the energy levels, the particles) that will make their appearance as excitations of the vacuum. In fact, physicists speak both of the mathematical functions or operators and of the physical processes that formally and actually "create" and "annihilate" a particle (a Something) in a vacuum.

OF NOTHING, SOMETHING, AND THE VACUUM: A STORY OF ORIGINS

In everyday life we are perhaps most aware of gravitational forces. We have jumped up and come down. We have been told about the tides on Earth and about the motions of the planets around the Sun. Now our most everyday experiences of real forces – such as the hardness of a rock we use as a convenient hammer or a bar magnet's keeping our refrigerator door closed – are due to electrical and magnetic interactions. Still, physicists argue that gravitational forces are rather modest and electromagnetic forces are not really so potent. For, they say, in these planetary and everyday experiences, what we observe depends on the addition of the effects of the very many molecules, 10^{23} and up, that we find in an ordinary piece of matter.[15]

Say, instead, as physicists do, we were to ask how strong is the force between two individual elementary particles. (Of course, this is the crucial move; it is just such a move that makes one a physicist rather than an engineer or an everyday layperson.) Then, one of the curious facts about this physicists' world is that the forces of Nature appear in a hierarchy of strengths – in stages, so to speak – rather sharply separated in magnitude (a fact employed in chapter 3 to explain "quite rarely"). The gravitational force between two protons in a nucleus is about forty orders of magnitude (that is, forty factors of ten) less potent than is their nuclear or "strong" force, and the electrical force is about an order of magnitude less potent than is their nuclear force.[16] Moreover, as the energy of the interacting particles becomes greater, the hierarchy seems to collapse, the forces becoming more equal in strength. Yet these higher energies are so large, so much greater than we see in the laboratory (but not in the cosmos), that we may ordinarily speak in everyday life of hierarchy and separation.

Given such hierarchy and separation, physical problems may often be treated as if each situation or stage were set up for the most part by all the more potent forces, while its special features can be understood by watching how the next-weaker forces disturb that setup – "perturbatively."[17] (So, for example, electrical effects are perturbations on the general structure of a nucleus, a structure mostly determined by the nuclear or strong interaction.) Or, the most potent forces define extended families of generic particles, such as the family or octet of "pseudoscalar mesons" (the pions, the kaons, and the eta), or, in a different vein, as in the families denominated by each of the types of neutrino (say, the electron-neutrino's family: electron, up-quark, and down-quark) – and the less potent forces effect the crucial differences among the particles within a family. (And, of course, a particle may be combined with certain others of its family or of different families to create more complex particles.)

Now, this story of a hierarchy of forces is just the one that is revealed in the Big Bang and in the cooling down of the universe subsequent to the Big Bang. Here, there will be three sorts of stage setting: the appearance of space and time and matter; the appearance of the various different kinds of forces; and the subsequent appearance of the variety of particles we know about every day in the physicist's laboratory and in the cosmic universe.

Initially – the Big Bang – the universe is said to transform from a state of very great symmetry to a state where there is spatial separation. (For our purposes here we need not say just how this "inflationary" process takes place.) Very soon after, it makes two transformations (about 10^{-35} seconds later, and 10^{-10} seconds after that): from a still highly symmetric world, in which all the above kinds of forces appear the same, to a state of greater orderliness (and the temperature and so the average energy of interaction is lower), when the different kinds of forces are now distinguished and exhibit very different strengths, and the kinds of Somethings, particles, are now rather more articulated. Also, as the universe cools down further, there will be a transition from a state where there is radiation (photons and neutrinos) but no massive matter such as protons, to a matter-filled, even less symmetric, more orderly world. (All of these transformations reflect the structure of "the standard model" and so-called grand unified theories and the forces at their most energetic.) Subsequently, and very roughly speaking, the more potent forces explicitly display themselves in actual comparatively stable heavy particles at earlier, more energetic, hotter times; while the less potent forces and lighter particles and more subtle differences must wait until things cool off sufficiently (due to the expansion of the universe) so that the general background heat does not overwhelm them.[18]

As the temperature declines, the various elementary particles – in effect representing different forces – sequentially "freeze out," becoming stable or resistant to vaporization into radiation by the mutual annihilation of a particle and its antiparticle: neutrons and protons representing the nuclear forces freezing out at ten thousand billion degrees Kelvin at a millionth of a second after the Big Bang, and electrons freezing out at roughly two thousand times cooler and one second later.[19]

Again, roughly, to be later in time is to be cooler is to be less symmetric and more orderly. Time and structure are intimately connected. And to be later in time is to have more degrees of freedom hidden, in the sense that they are frozen out into particles, held in place by the particles' walls, the ambient thermal energy too low to free them from their bindings. What impresses us now, much later in time, is the orderliness of the leftover degrees of freedom, whether it be in a crystal or in the family structure of the elementary particles.

Again, in this story, Nothing is comparative; it is what existed before there was the appearance of the vacuum's orderliness and Something within it – at each of the various transformations and stages, where Something might be matter itself or the later separation of matter into comparatively stable particles of different sorts. Something appears as a result of an abrupt transformation. It is a transformation that becomes possible once the system descends to or goes below a temperature threshold such as a freezing point, and the transformation is perhaps triggered by an arbitrary event.[20] Once an orderly Something can appear, there is a continuous process (in temperature, and in time in the case of the Big Bang) which allows for various amounts of that Something to exist – usually less of it as things cool down further since there is less energy available. Such a sequence of transformations may be said to be hierarchical and self-consuming – the Something of one stage becoming the Nothing of the next – each transformation articulating the previous one. And the hierarchy may be seen as a product of time, a narrative unfolding as we tell the story of the Big Bang. The unfolding creates a more orderly world, more particular and less symmetrical. In this story the systematics of structure and order and narrative-temporal history are to be made identical.[21]

Hegel (1807) might well have approved of this kind of account, for it is in his terms dialectical, and rational and historical – a systematic hierarchical self-articulating structure that develops and displays itself in time.[22] And Aristotle might have spoken of drama and of turning points in that drama, when the possible becomes actual. What is called for here is a *rhetoric of theater:* the creation of a suitably empty stage, the continuous development of appropriate characters upon it, new stages and hierarchical unfoldings, and events that trigger major transformations.

Moreover, such theater is educational, so to speak, as well as scientific and dramatic. For cultures become known to themselves and to others by the stories they tell, and by their modes of reenactment of their origins and their cosmologies. And I have just been telling such a story.

Now in the Big Bang cosmology, the universe might be taken to be an expanding gas of matter and field and energy and spacetime. As such, it is cooling down (for that is what expanding gases do, as in a refrigerator). And the various transformations are analogized to well-known behaviors such as freezing. So the Big Bang history is analogized to the corresponding behavior of well-known substances, such as an ordinary gas or liquid or magnet. Hence, we have here what might be called an *analogy of substance*. For physicists, the world is always like something we already know quite well, something material and everyday. The macrocosm as we might conceive of it is a mirror of some quite mundane microcosm we already know, or perhaps the analogy goes the other way around.

Finally, physicists account for the appearance of something from nothing by employing models of *discontinuous* transformations as their pictures of Nature's dynamics. These models naturally allow for qualitative transformation without the need to provide, at least manifestly, a continuous connection between something and nothing (and hence the especial attractiveness of transformations that are spontaneous as well). So when physicists take Nothing to be a relatively greater arbitrariness and symmetry, then Something need be just what is in accord with an orderliness that is sufficiently more restricted than and so incommensurable with Nothing's symmetry.[23] And in this manner something arises from nothing.

Just as the separation of light from darkness is the foundation for the Creation in Genesis, the very great conceptual achievements here are the set stages, often in hierarchy, namely, the sequence of orderly but empty vacua – where, again, each vacuum and its Somethings, each stage, each act in the script, may become a Nothing with respect to its successor vacuum and its Somethings.

Each vacuum is a black stage with the potential for an orderly Something to happen upon it. And each black stage has much device and degrees of freedom hidden behind its orderliness – but we can get away with ignoring that device and those degrees of freedom as we watch the show. To be sure, there are fluctuations, much as there can be melted fluid-like bits or inclusions in a solid or moments of breaking face or backstage accidents in theater. These allow hidden degrees of freedom to show themselves. But in a good vacuum those fluctuations are tempered and comparatively unimportant, and we may treat them as random and incoherent events. However, at transition points or in high drama they are of overwhelming significance.

I want to reiterate that physicists do not actually see emptiness – a vacuum as such. For that makes no practical sense. The temperature is always greater than absolute zero, physicists say, and so there is thermal excitation – namely, Somethings. For example, we always encounter a crystal vibrating a bit (and so it has Something in it). More generally, we always see the stage lit a bit, for we can only see by the light of a probe (a flashlight), which again in effect heats up the stage a bit.[24]

So we can never literally see that unoccupied unlit stage. And the time at creation is similarly inaccessible to us.[25] A vacuum, as such, is on the edge, in-between.[26] It is *liminality* itself.[27] What we actually see is a world we take provisionally as orderly, and then we abstract away from that by dividing the world into a stage and the actors upon it: namely, an orderly emptiness or vacuum, and Somethings in accord with that orderliness that populate that emptiness.[28] What is Genesis, or the Big Bang story, but such a marvelous set of abstractions?

To recapitulate: The rhetoric of theater makes it natural that there be hierarchy, unfolding, order, and arbitrary triggerings. The analogy of substance is a concrete physical situation on which to model other physical situations, so making them familiar and natural. And the account of discontinuity shows how a multitude of degrees of freedom may be well tamed and so they are hidden in a coherent fashion; it allows for the abrupt appearance of the orderliness of the vacuum, so permitting exhibitions of that orderliness, namely Somethings; and, as we shall see, it permits well-isolated disorderly phenomena which we may count as transient Nothings (fluctuations). Setting up a stage hides much of ordinary life, yet then there is the possibility of theater and action upon that stage.

III

SETTING UP THE STAGE: THE ACTUAL CRAFTWORK

Physicists want to invent, discover, or experimentally set up a smooth and regular vacuum, so hiding lots of degrees of freedom and providing a nice orderly arrangement for some of the leftovers. They also want that stage or vacuum to be stable to mild insult and poking. And then there will be both an empty-enough stage and one not too readily destroyed by what happens to take place upon it.

Now such a functional and rhetorical definition of the vacuum does not say in detail what Something is, or how we might go about finding or inventing a good vacuum. Experimentally and conceptually, setting up that empty stage on which Nature's appearance makes theatrical sense is for the physicist a detailed technical problem. Moreover, as we might expect, the what and how of a good vacuum are defined by conventional methods, particular models and procedures that have worked in the past and have come to be justified in principle. The analogy of substance (that is, to a specific substance) is crucial to the craft of science, just as a tradition of actual presentations is crucial to theater. And, in particular, the experimental setups and the mathematical technology employed by physicists to study phase transitions and equilibrium are extraordinarily influential as models for setting up a vacuum. New ways of achieving a good vacuum tend to be reinterpreted so that they seem naturally about phase transitions and equilibrium, namely, these paradigmatic conventions.[29] These methods and archetypes are more than influential, they are taken as necessary, as "it must go this way." Along the way, the intuitive notion of a vacuum is generalized, and that pumped-out empty place comes to be taken as any place in which there is a pervasive orderliness.

Let me preview the rest of the argument of this part. I first review a series of conventional technical rules and physical constraints that are often employed to produce a good vacuum and the generic principles that are used to justify each constraint. Then I discuss what it means to be a good enough vacuum, and how such a vacuum does more than we might expect: Degrees of freedom are tamed, an emptiness is created, a new orderliness is found, and some formal procedure (formal mathematics, or formal analogy to another situation) does all of this work automatically. A countertheme in these practices is how the stage is set as a continuous incremental process. But then there needs to be a rule to pick out the one true vacuum out of this continuum, so restoring the discontinuity (or at least the markedness) of the moment of creation. Again, it should be noted that even this rhetoric never will prescribe Nature's stages; it only prescribes their form, how the world must appear. (And, of course, it might not even appear these ways.) And, in any case, we need to have the answers in mind, actual experience of the natural world – in fact, we always had them – when we interpret the rhetoric so that we read off its implications correctly.

IDEOLOGIES FOR A VACUUM: CONVENTIONS FOR TAMING THE DEGREES OF FREEDOM

Our purpose here is to describe how physicists' practices become sacralized. "The way the world must go" is in effect an ideology, one that offers what are taken as deeper reasons or principles to justify as necessary and even as a sacred providence the practical rules or physical constraints (for example, constant temperature) employed to set up a vacuum. I should note that this section and the next one are moderately technical and quite detailed, but only in their examples and not in their main point – that physical constraints are associated with justifying ideologies, the constraints and ideologies together leading to a good vacuum, and that there is a theoretically productive dialectic between continuity and discontinuity in defining that good vacuum.

Ordinarily, we pump down to get a good vacuum. A vacuum pump gets hold of errant molecules – in effect, spare degrees of freedom – and sweeps them away. Analogously, if we cool down a system, many of its degrees of freedom may also be swept away or at least tamed, some so rarely excited we may ignore them, others literally becoming frozen out – into a solid having an orderly substrate of crystalline structure as in a metal or ice, or into long-range magnetic alignment of its atoms as in a permanent magnet. Thus, many degrees of freedom are swept away or tamed by the pump and the refrigerator and other constraining devices (often called experimental setups) – emptying the stage. And those degrees of freedom that are then left might show themselves (as orderly Somethings) in a particularly simple and understandable way. The resulting system or world is *constrained:* to be in thermal equilibrium, as is the gas in a balloon; or of fixed dimensions and rigid, as is a solid; or repetitive, as is a crystal; or aligned, as is a magnet.[30] Also, that orderly thermal equilibrium, rigidity, repetition, or alignment allows for small disturbances of that orderliness (Somethings!) – such as sound waves in a crystal, taken as modulation in the repetition. (Note that different constraints are sometimes apparently equivalent in their effect, so that rigidity and repetitiveness both characterize crystalline solids.[31])

Constraints impose themselves upon and so tame many of the degrees of freedom, and in effect the constraints create an orderly vacuum. (Of course, again, it is just in the actual experimental setup or theoretical conceptualization that the physicist does the constraining.) Now if we have a conventionally set theatrical stage in mind, such a confluence of orderliness and taming may seem natural, almost a matter of definition. But that theatrical analogy in itself does not justify the physical constraint. Physically, justification is usually offered in terms of a deeper technical feature that physicists say characterizes Nature most fundamentally: the natural world is ergodic, or exhibits broken symmetry, or is invariant, or is causal or correlated – that is why it is in equilibrium, rigid, repetitive, or aligned, respectively.

Technically, a world is said to be ergodic if, roughly, every allowed detailed arrangement of its components will happen (say with equal likelihood, as in the plenitude of chapter 3, or weighted by a statistical factor, $e^{-E/kT}$, if there is noise). And, then, in many cases thermal equilibrium of that world turns out to be the most likely situation, since thermal equilibrium represents most of the possible different detailed arrangements. What are hidden are the less likely (and in fact heterogeneous and in disequilibrium) possibilities, since they are overwhelmed in the number of detailed arrangements they represent by the much more likely, apparently uniform one.

A world exhibits broken symmetry – that fallen pencil – if there could be, say, a preferred although otherwise arbitrary direction in an unbiased space, and if such orderliness is more stable than is indeterminateness. And a rigid solid surely provides for such stable orderliness.[32] For to change even just a bit of that orderliness, there being say a bit of melting in a small region of the solid, is very difficult, since the rest of the solid is overwhelmingly influential on that region and so the bit of melt will rapidly recrystallize. What is hidden are the arrangements of the atoms that produce less orderly, more fluid, states.

A world is said to be invariant if it looks the same after a particular kind of transformation – and repetition is wonderfully invariant. What is hidden is what is not invariant; its lack of repetitiveness dooms it to unrecognizability. A world is taken to be causal if we can predict what will happen nearby (in space and in time) from the present situation, and alignment is manifestly quite correlated and so in this sense it is causal. And here what is hidden is what is not correlated, since it is not so closely connected to what is nearby, and so it will be less likely to be noticed.

Deeper features are deliberately built into the formal and mathematical structure of physical theories or are found latterly to be features of those theories, in each case seen as necessary if those theories are to make sense and be "physical." (Recall our description at the end of chapter 1 of the deeper features of electromagnetic or gravitational theory.) But most crucially for our purposes here, deeper features stand in for their practical vacuum-making effects, effects which are now seen as concrete realizations of the features' deeper theoretical necessity. When ergodicity, broken symmetry, invariance, and causality are said to reflect deeper facts about Nature, physicists are also saying that Nature *should* be in thermal equilibrium, rigid, repetitive, or aligned – at least mostly – if it is to be of concern to physicists.[33] And so Nature will exhibit a good conventional vacuum – namely, the gas in equilibrium, the solid, the crystal, the magnet – as well as orderly objects (Somethings) in that vacuum. These are objects that represent small disturbances of that uniformity, broken symmetry, repetition, and alignment, such as elastic vibrations of that solid. As noted above, different principles (and their constraints) often will have the same effect, and so what we see may be justified in more than one way. We might say that Nature is overdetermined. We then might try to delimit those realms in which such similarity or overdetermination reflects even deeper features. And so forth.

Some other conventional physical constraints – besides thermal equilibrium, rigidity, repetition, and alignment – are smoothness, "bell-shapedness," and insusceptibility or insensitivity to external influences, which are justified by saying that Nature is in principle a continuum, or it is statistical, or it is optimal, respectively. Here, Nature is constrained to be smooth as in a flow, or well characterized by a mean and a standard deviation as is a bell-shaped distribution of a random variable, or in mechanical equilibrium as is a spring under tension.

Technically, a world is said to be a continuum if limits, like derivatives, are well defined.[34] In the limit of smaller and smaller incremental changes, a derivative or a velocity is always found to be the same no matter how we measure it. Or, consider the "continuum limit" of a crystalline solid: While the crystal is made up of discrete atoms, it exhibits vibrations as if it were a smooth continuum, in the limit of wavelengths that are long compared to the distance between individual atoms. (More generally, the behavior of many of its properties does not depend sensitively on the details of its microscopic structure.) Such limits and derivatives then allow for dependence and differential equations, the most powerful account of incremental and so smooth change.

A world is said to be statistical if it has many independent parts (like many coin tosses). The collective effect of those parts, almost no matter how peculiar the parts are individually, acts much as the proverbial sum of random errors, or, technically, a sum of identical independent mean-zero random variables – namely, a bell-shaped curve, fully characterized by its mean and standard deviation. All sorts of other distributions and fluctuations, namely other degrees of freedom, are proscribed away by theoretical fiat (namely, the central limit theorem in all its varieties).[35]

And, finally, a world is optimal if it is insusceptible or at least insensitive to external influences – since optimality is defined by the maximum value of a criterion, let us say, and in general *at* such extremes (the top of a mountain) small changes have small effects. (This is, of course, one of the main messages of the calculus.) And so the handles onto all sorts of what we take as internal complexities and their degrees of freedom will be comparatively untouched by our probes, and in effect those degrees of freedom are damped and so hidden.[36]

Each constraint is taken as a technical criterion for a good vacuum, its associated deeper feature bringing in the constraint by the back door, so to speak, the deeper justifying feature being an ideological and aesthetic commitment to how the world must go, the core values for a physicist. Empirically, the world just might not go in accord with a particular constraint and deeper feature; but physicists work hard to find an interpretation of the constraint and of the feature so they are fulfilled. The commitments to rules of orderliness – uniformity, rigidity, repetition, alignment, smoothness, bell-shapedness, and insusceptibility – are not abstract commitments.[37] They have to deliver on their promises, or be made to so deliver. They must tame many degrees of freedom so there really can be an orderliness, one that reflects a deeper feature of Nature. If the Nothing is properly swept up, there ought to be left over a nice vacuum with Something in it.

THE DIALECTIC OF FINDING A GOOD VACUUM: A PHENOMENOLOGY

As we discuss in chapter 5, a good vacuum allows physicists to observe phenomena in terms of that vacuum's orderly Somethings, Somethings which are simple and recognizable. For when one pokes properly at a vacuum, with the right kind and level of intrusiveness and gentleness, the response is just those Somethings, the rest of the degrees of freedom still being tamed and hidden and held in. Conversely, that the effects of poking are simple and reproducible is taken as a sign of our having achieved a good enough vacuum for this particular physical situation. (That a piece of theater works well suggests we have set the stage properly.) Now, effects will be reproducible if we have swept up enough degrees of freedom so that we have sufficient control over all the rest – for otherwise, untended-to degrees of freedom could arbitrarily affect what we see. (We do not want strangers wandering onto the set.) And if reproducible, the disturbances of the vacuum will then be seen as "effects," as consequences of our own interactions with the world. (In classical theater, we want a sense of causation and agency.) Observation now becomes possible because we know how to generate robust phenomena and what kind of signals to look for.[38]

Now a good orderly vacuum is, in fact, overdetermined in that the actual achievement of any one of the goals seems to give the physicist the rest by the way. To repeat: Degrees of freedom are tamed, an emptiness is created, a new orderliness is found, and some formal procedure (formal mathematics, or a formal analogy to another situation) or some experimental setup does all of this work automatically. These are both grammatical and practical demands. Taming degrees of freedom means there is emptiness, which means there is a residual orderliness, and this work is done by formal sleights of hand or through the design of experiments. However, the experience of the physicist is rather more contingent, the work of creating or setting up or conceiving of the vacuum rather more subtle, than any grammatical identities would imply. And that wondrous overdetermination is only apparent after a good vacuum is found. A good vacuum is in essence those right degrees of freedom that Weinberg speaks of. The commitments to the various constraints and their deeper features might be likened to a faith, their eventual automatic effectiveness being seen as its miraculous confirmation.

Now, in the end, the story of constraints and deeper features proves to be insufficiently satisfying. For example, that an optimization principle works does not say just how and why it works.[39] So scientists also demand an account of how orderliness sets in, just how the sweeping up of the irrelevant degrees of freedom takes place, just how the constraints are actually fulfilled. Such a mechanism is taken as the sign of a real explanation. Note as well that such an account is ironic in its effect, since creation is no longer a marked moment or a sharp transformation but now is shown to be a process, a process that often proves to be incremental.

As an example, consider two mechanisms accounting for the freezing of a fluid. Freezing may be thought to set in either as a fluctuation in an unstable system or as a matter of degree of orderliness. Close to or at the freezing point, small regions of the fluid fluctuate between the liquid and the solid phases. Those fluctuations, say toward solidification, can be large enough to grow rather than be damped, especially if we are below the freezing point, and so they might encompass the whole fluid. Alternatively, we might think that a "hot" crystal about to melt is like a "cool" liquid about to freeze, and so orderliness might be said to set in gradually and more and more pervasively.[40]

Or, consider our other recurrent example. (See Figure P.1.) Near but above the transition point (about a thousand degrees temperature, Kelvin), small regions of a bar of iron fluctuate to and from the aligned magnetized state. Here the patterns of fluctuation look the same at a wide range of magnifications or scales. Now say we assume that this "scaling symmetry" turns on continuously with temperature (see chapter 5). Domains, taken as regionally coherent collections of aligned atoms, are present for a larger and larger range of distances or scales as the temperature approaches the critical temperature (say from above the transition point). Eventually, at that transition temperature, the whole piece of iron is coherently involved. The distance beyond which atoms tend not to be correlated – the correlation length – is in effect infinite (at least compared to the distance between neighboring atoms). And so bulk permanent magnetization and its orderliness is now possible, its particular direction arbitrary or perhaps a consequence of the influence of an external field. Put differently, the notion of "bulk" makes sense only when correlation lengths are suitably large, namely when they are macroscopic (or "infinite"). In this analysis, degrees of freedom are said to be successively "thinned out," as every second, fourth, eighth, sixteenth, . . . atom might be much more likely to be lined up with its neighbors (and so it is less free) as the temperature gets closer to the critical temperature.

Rather than having a set of constraints that simply apply, they are to apply in degree (again, as one gets closer to the critical temperature). Now, physicists speak of an "order parameter" measuring the degree of orderliness, such as a measure of magnetization or of spatial correlation. What is still discontinuous in this story is the actual appearance of a qualitatively new property (that pervasive or bulk orderliness), even in an infinitesimal amount, which occurs only *at* the transition point. Such a discontinuous *and* incremental appearance of an orderliness (or disappearance of a symmetry) is not at all peculiar if it is represented by a graph that rises "infinitely" fast, say as we approach the transition from above. Put differently, the appearance of even a bit of pervasive orderliness pollutes a symmetry and, physicists say, "lowers" that symmetry to greater orderliness.

Again, in each of these cases, the new vacuum or orderliness is taken to be quantitatively and incrementally connected to the Nothing from which it is derived, even though its most characteristic feature is its pervasive orderliness, its qualitative difference from that Nothing. And the amount of the qualitatively different Something in that new vacuum may be changed incrementally, but the actual appearance of the vacuum, as such – that orderliness – is abrupt and discontinuous.[41] It would seem that for physics as a practical theoretical enterprise, and for Nature, the dialectic of continuity and discontinuity, and of quality and quantity, is actually productive and is not to be resolved.[42]

Even if an orderly vacuum and the more symmetric Nothing from which that vacuum arose are somehow to be made conceptually comparable and connected to each other, Nature itself is, in fact, not in general so undecided – except, of course, at the transition point, which is filled with fluctuations.[43] Put differently, for situations that physicists take as interesting, it seems there must be only one stage, only one kind of vacuum in effect at any time.[44] (Of course, there are in Nature other more mixed situations, but these are not usually taken as physically interesting vacua. They are too messy, so to speak.) There is either Nothing's symmetry (which, of course, is a vacuum itself) or the new vacuum's order, not both: water or ice. In the end, some qualitative differences really are incompatible. So J. Willard Gibbs's phase rule (1876–78) limits how many phases of matter can coexist, depending on the number of independent degrees of freedom. And Henry Adams wrote about this physical law in hyperbolic terms ("The Rule of Phase Applied to History") just because it gave an account of why there might be a systematic hierarchy of qualitative changes in history and in thought.[45]

Physicists employ generic principles to pick out the one true vacuum, for a particular set of external conditions such as temperature and pressure, from among candidates which are supposedly only incrementally different from each other – in effect, a dialectical response to the accounts of *how* the stage is set. Typically, a principle picks out the optimal candidate, the one that has the most or the least of something, whether it be entropy (ways of being rearranged), or energy, or volume in a mathematical space.[46] And so, as we have seen already for deeper features, such principles justify that vacuum and its orderliness by gracing it with a property – optimality of a particular sort – which physicists have come to recognize from their experiences and justifications as universal and sacred, a property that as we shall suggest guarantees as well that the vacuum is both empty and stable.

For example, if the true vacuum is taken to be the situation with the least energy, and if there is only one such minimal energy state, we have picked out a unique vacuum – the lowest or ground state. It is empty just because, as the lowest energy state, no energy is available for Somethings to appear (but see the next section, on fluctuations); and it is stable because there is no lower energy state to go to. If the vacuum is taken to be the most rearrangeable situation – namely, the state of greatest entropy – then again, if there is only one such situation, we have picked out a unique state (such as equilibrium for a closed system). It is a good vacuum because, if the rearrangements of its components are likely to have no effect, it will exhibit few if any disturbances. So it will be empty. And it is stable, because rearrangements that do have an effect are quickly drowned out by the ones that do not. (It is assumed, and this needs to be shown in order to justify the principle, that the most rearrangeable situation is not only a plurality but an overwhelming majority, so that it includes most of the conceivable arrangements of the components.)

In sum, a good vacuum – orderly and stable, with nicely defined Somethings, a vacuum within which one can do physics – is defined in practice by demanding that the constraints or their associated deeper features be fulfilled. If there is more than one situation that fulfills a principle – what is called degeneracy – then the principle has not done its work.[47] There are, it seems, some degrees of freedom which are on the loose, out of control. For physicists such degenerate situations must have some differentiating characteristics. What is crucial is to find just one vacuum, empty and stable – and there must be a way of doing so. And so degeneracy leads to a search for a sharper definition of a vacuum's orderliness.

Put differently and recalling chapter 3, in cultural terms, the lack of a distinction when we believe there ought to be one leads to degeneracy, considered as a pollution, a mixing of what should be separate. Degeneracy is resolved when suitably maintainable distinctions (physical forces or interactions) are found and turned on.

THE ANALOGY OF SUBSTANCE, ONCE MORE

These formal rules for finding the one true vacuum for a situation are just that, formal rules. They are rules employed as everyday of-courses by a physicist. But they, alone, tell us too little about our universe to be definitive. And even if they are fulfilled, they do not guarantee that we are choosing the right good vacuum, one in which we are employing the right degrees of freedom. As we have said already, the specific pictures and details matter enormously. For example, the great physicist Paul Dirac's picture (1930) of an electromagnetic vacuum as a just-filled-up "sea" of hidden negative energy electrons – the energy level being analogized to the distance above or below sea level – might be shown to fulfill the rules for a good vacuum. This vacuum can be seen to be empty and those electrons be hidden much as a rare gas such as neon is not interactive, and hence empty, because its atomic shells are just-filled. But, despite fulfilling all the rules, one needs to then invent or discover or set up such a vacuum first (and, for example, to note such an analogy to the rare gases) – and such an invention is a very great achievement. The rules regulate that invention but do not prescribe it. Similarly, it is virtually impossible to know the crystal structure of ice before actually seeing some frozen water, even if we know what must happen in principle.[48]

Even the formal abstract mathematical techniques and the actual experimental setups used to do much of this work of finding and characterizing an orderly vacuum are employed in quite particular and conventionalized ways. If averaging generates smooth situations, and mathematical symmetry rules (or groups) characterize repetitions, particular ways of averaging and particular mathematical groups are employed again and again. We need not judge whether this is either because physicists are conservative in their commitments or because the world is rather simple and is recurrently structured in the same ways. In actuality, practical effectiveness would make anyone, including physicists, be committed to particular mathematical techniques which were successful again and again – and those techniques might then be justified as the right techniques either on empirical-practical grounds or on more transcendent ones.[49] So it comes to make practical, ideological, and theoretical sense to see if, with modifications, a previously successful and justified technique, used in a particular way, will continue to work. For example, from physical argument tested by past experience we might take it that an infinite value of a function indicates a point at which an orderly vacuum sets in. But what if at first for a new situation the mathematical technique does not pick out a vacuum? It would not be unusual to say that perhaps the mathematical idea is correct and the formulation of the physical problem is wrong.

To have mastered the various rules and their associated obsessions and imperatives, as I have been describing them in these chapters, is not enough to make you a physicist – although if you are a physicist, you will share those obsessions, you will share in that subculture's necessary way things must be. Rather, you become a physicist by your involvement with the *particular* problems and material physicists take as interesting, and along the way you pick up the rules of play. The obsessions and rules are rarely argued for explicitly. They are taken as of-courses, and as such reveal foundations of the enterprise of doing physics. And other physicists could not make sense of what you did if you did not follow these rules, more or less, in the context of a canon of particular problems.

IV

Even when the vacuum is supposedly empty, orderly theatrical events continue to occur there. This stage is not to be hermetically sealed, at all. Visitors from other stages drop in regularly. And the stage is always lit a bit.

FLUCTUATIONS IN A VACUUM

If a good vacuum has been found, then the leftover degrees of freedom express themselves simply and understandably. There are not too many of them, and they are recognizably orderly. Again, at this point we might fantasize that we could actually clear the stage completely, and contemplate an empty vacuum, an abstraction made real – truly empty space or that simple orderly crystal. But for Nature there never is just an empty stage. Some degrees of freedom are unavoidably active as Something. The actual vacuum (or, as physicists say, the physical vacuum) exhibits *fluctuations:* whether it be liquid droplets in a solid or sound vibrations (or phonons) in that crystal; whether it be, as part of the elementary particle physicist's physical vacuum, the appearance of transient electron-positron pairs ("vacuum polarization") and transient radioactive decays ("virtual transitions"); or whether it be the vibratory motion of a particle that has no external forces acting on it ("zero point motion").[50] And these parenthesized oxymoronic names for these unavoidable degrees of freedom really are ironic. For if one could truly have an empty vacuum, a vacuum with no fluctuations, these phenomena are either supposedly hidden (that is, they are Nothing) or they are yet to appear as simple disturbances of the orderliness (that is, they are Something).

Moreover, fluctuations prove to be remarkably sensitive probes of the shape of the world. They get at its edge, so to speak. Technically, physicists say that if the average energy available from the external world is large (say the temperature is high enough) compared to the energy that is needed to excite a degree of freedom at all, then that degree of freedom makes its full-blown appearance; and if the average energy is comparatively small, then that degree of freedom remains silent or hidden or frozen out.[51] (A crucial idea is that there is a minimum energy of excitation, the quantum; this is the great contribution of Planck and Einstein.) The hard part, and what is taken as the interesting and detailed physics, lies at the edge, when the external and internal energies are roughly equal, and so comparatively small differences in those energies, namely, fluctuations, turn out to be the dominant phenomena.[52]

True to form, physicists try to tame fluctuations. First of all, as we have seen in chapter 1, it is argued that fluctuations are ineliminable. So the irony is turned back onto itself. That is just what the Heisenberg uncertainty principle or the second and third laws of thermodynamics do. The Heisenberg principle says that even in a space that we take as empty, there must be fluctuations in the number of particles or, correspondingly, in the values of a field (each supposedly equal to zero). And the second and third laws may be shown to say that at any temperature above absolute zero there must be fluctuations, in the energy for example.[53]

That fluctuations are ineliminable does not mean physicists do not tame them. In practice, what a physicist means by such an unavoidably active degree of freedom is that there is no way to firmly get hold of just that degree of freedom, or it has not yet been frozen out, and there is no means of controlling or predicting any single one of its appearances. We are to be unavoidably ignorant of the particular features of each prospective fluctuation. That is just why these phenomena are called fluctuations (rather than discrete individual events). For a physicist, that ignorance means that each fluctuation is, like successive coin tosses, more or less statistically independent of the others. Knowing how one fluctuation turns out tells us nothing about any particular predecessor or successor. And probability and statistical methods are the way of expressing our knowledge of that ignorance; namely, fluctuations are distributed randomly, say in space, time, and intensity, often according to a bell-shaped curve. So, again, an orderliness is found, even if the vacuum is not empty and never can be.

Now the sharp discontinuity of a vacuum's onset is seemingly challenged by fluctuations, which, near or at the transition point, would seem to intermix the kinds of incommensurable orderliness. But each one of those individual fluctuations, representing our detailed ignorance, cannot give us any information about which side of that point we are actually on. Rather, we may only draw inferences from the distribution or spectrum of fluctuations, the range of our ignorance.

All of this is, to be sure, a curious form of theater. I bring it up to emphasize that fluctuations represent an ordered ignorance about marginally held-in degrees of freedom. Not only does the physicist's stage setting or vacuum demand a rhetoric of theater, and an analogy of substance, and an account of discontinuity, but there is as well an ontology of liminality, where some of the natural objects are those that are revealed at the edge, on the limb, when the relationship of each fluctuation to the world is as unknowable as can be.

V

ANNEALING THE WORLD: MELTING IT DOWN AND SETTING A STAGE

With good experimental and theoretical reason, physicists have come to believe that for many a complex system, the effect of the constraints, shielding, and orderliness that radically restrict a system's degrees of freedom and give that system a pervasive coherence may often be epitomized by a temperature – a measure of the average energy per active degree of freedom of the system.[54] Moreover, in general the number of active degrees of freedom rises slowly if at all with temperature; but, at what are called transition points or thresholds there will be a rapid change in the kind and perhaps the number of active degrees of freedom. And as the temperature rises further, each active degree of freedom becomes even more active. Those that are still in cold storage make their presence felt through even larger fluctuations.[55]

Heating things up, as we get closer to a point of phase transition, Nature is no longer so well characterized by its former orderliness and demographics. But we cannot yet use the good degrees of freedom which will get hold of that next, even more symmetric, phase. The world is still too restricted to be so free. The fluctuations are now not so damped, not restricted to being so small in size. And probes, rather than gently – or say a bit less than gently – disturbing the system, can have dramatic seemingly infinite effects.[56] For example, it becomes acutely easy to change the direction of magnetization of a permanent magnet (although the amount of magnetization itself is quite small). Eventually, if further heat is applied, the temperature rises, the stage melts, and there emerges a less orderly, more symmetric system.

The cosmological story of the origin and cooling down of the universe and the mundane story of freezing are the same story – namely: an account of Nothing, of creation, and of an orderly vacuum and Something within it, each stage being governed by principles or constraints which make for a good vacuum; an account of fluctuations, just that ignorance we must have about the empty vacuum; and, finally, an implicit commitment to a hierarchy of forces, denominated by an energy measure or a temperature, so that there are distinct stages, and at each stage we may for the most part treat weaker forces as perturbations.[57] These stories exhibit a rhetoric of theater, an analogy of substance, an account of discontinuity, and an ontology of liminality.

Again, each story is a creation story, telling how something might arise – discontinuously or abruptly – out of nothing; and in this sense, both the cosmological and the mundane stories are theological. And, as in Genesis, creation leads to an account of the orderliness of Nature. What is preserved, always, is the conceptual separation of matter or radiation from an orderliness which they represent, of Something from the vacuum, of particles from empty space, of excitation from ground state. The great obsession is to create a stage on which observable phenomena, these excitations, appear as such. Can we get just the right degrees of freedom on stage? Can we set up a society in which a modern economy is possible?

But getting the right degrees of freedom is just what physicists are trying to do in the division of labor, or when they find the right parts, or when they look for rules of forbiddenness. We have yet to talk much about the physicist's actual practical way of setting up the world so that it is observable and knowable, and of the physicist's corresponding motives and obsessions. And it is to that way we now turn.

FIVE

Handles, Probes, and Tools: A Rhetoric of Nature

A Craft of Science; Some Handles onto the World (Particles, Crystals, Gasses; Analogy; Phase Transitions; Knowledge Is Handling). Probes; Objectivity and Inelasticity; Probes and Handles. Tools and Toolkits; A Physicist's Toolkit; So Far.

THE ARGUMENT IS: PHILOSOPHY HAS OFTEN PROJECTED A VISUAL analogy of knowledge – the knower as spectator – into its discussions of science. But physicists speak of what they are doing in terms of an Archimedean, haptic, and instrumental analogy. They sensitively get hold of the world (the Archimedean fulcrum) and so get a feel for it; and they craft explanations by employing as concepts just those instruments or tools they use to take hold of the world. Knowledge is handling, and Kantian transcendental conditions are actual experimental setups and theoretical models. The Archimedean analogy not only describes the physicist's research work itself, but also the physicist's theoretical structures – *handles* being degrees of freedom, *probing* modeling our interaction with Nature, and *tools* often being physical models and mathematics as well as experimental equipment. In earlier chapters we have been describing some components of such a toolkit. Here we review those descriptions in terms of handles, probes, and tools. Then we explore the technical and rhetorical structure of the toolkit, both in terms of mathematical and diagrammatic tools and in terms of a rhetoric for addressing Nature. The story of craft and handles is a commonplace for the physicist. It is the way physicists describe their work, just how it is actually done, as Maxwell described (p. xi).

I

A CRAFT OF SCIENCE

The stories physicists tell about the world, whether they be in terms of manufacture, mechanism, kinship, or creation, are about the way the world *is*. So the world is not only *like* a mechanism, it *is* a mechanism – at least provisionally, as long as you can get away with it. Professionals and craftspersons act this way, taking their arena for action as real, fully itself, even if it is shown to them that they are "merely" practicing modes of analogy and metaphor and rhetoric. Often, they are aware of this representational fact, but as long as their stories work in practice, what does such a showing-up mean? How would they do without analogy?[1] Whatever else – and I am not at all precluding reflection on the work itself – the world is (or goes) *this* way.

Now physicists employ their models and metaphors in rather particular and idiosyncratic ways. In training neophytes, the major effort is to get the student not only to see the world in terms of a particular model, but to take that model in a particular light, associating with it particular kinds of detail, relevant features, and ways of acting. To ask the hue of an electron is not to appreciate what is crucial about particles conceived of as billiard balls (even if we of late have attributed something called color to some elementary particles). So, saying how the world "must" go invokes not only the models and metaphors, but the traditions of their use within physics as well as their traditions within the general culture. When Nature is likened to the division of labor, that likening is within a particular interpretive tradition of that notion.[2] As I have just said, to accuse the physicist of employing everyday models and metaphors is no more interesting to the physicist than accusing the novelist of writing fictions and employing genre conventions. For it is almost always in the details and how they are marshaled that Nature's resistance to our conceptual powers shows itself, and so we may learn something surprising about the world.[3]

Now implicit in each of the stories I have been telling is an account of how we actually get at the world, whether it be by means of particles and fields, differential equations, classification rules, or orderliness. But so far, the accounts have been remarkably passive, as if the world were just there, not much having been said about how we actually learned about it or got at it.[4]

For a physicist, and I suspect for many other inquirers, observation and knowledge are actions. (1) We know about what we can get a *handle* onto – handles which we can use to bring some thing along, shake it, or pull it apart. Knowledge through the handles, and theorizing about those handles, is all the knowledge we shall ever have. To emphasize the instrumental character of degrees of freedom, we say that the right degrees of freedom are good handles. (2) We observe by *probing,* by ballistically scattering a probe off an object, by so taking hold of a handle, shaking or "exciting" it, paying attention to how the world responds, and noting those responses in terms of phenomena. And often these phenomena are actual objects, like particles, stimulated or created by that probing. (3) And we theorize, so creating systematic knowledge and explanations, using a *toolkit* of models and stories: a set of instruments employed in a craftsmanlike way to put things together into a nicely finished whole, along the way inventing better handles and better modes of probing.

So, for example, temperature turns out to be a good handle onto a steam engine, among other things telling how efficient such an engine can be. Heating up the steam, putting some heat into that engine, is a way of probing its workings by finding out how that temperature changes. And thermodynamics provides some lovely models to be employed as instruments or tools, such as Carnot cycles, for theorizing about those workings and for justifying why temperature is a right degree of freedom. Or, consider a baking potato. We observe the potato by probing it with a fork, noting how difficult it is to push in the fork and the wetness of the tines when the fork is withdrawn. Here, the handle that the fork gets hold of is the hardness and dampness of the potato, and our model is a notion of cooking: that as something is more cooked, it will, at least at first, become softer and drier. Or, consider observing a nucleus by shining highly energetic light onto it and watching how the nucleus changes, perhaps even its then emitting particles as a consequence of its having become radioactive in that interaction: Photon + Nucleus → Altered-Nucleus + $X + Y + \ldots$ The photon particles, which compose that light, probe the nucleus, perhaps penetrating its surface, so getting hold of one of the nucleus's handles. And the photon shakes those handles, exciting as well some of the nucleus's otherwise hidden or internal degrees of freedom. The absorbed photons, the altered nucleus, and the emitted particles (*X*, *Y*, . . .) we see in response to our probing tell us about the original nucleus's properties, such as its spin and shape and symmetries: namely, both those handles attached to the nucleus that the photons take hold of directly, and the nucleus's internal degrees of freedom indirectly shaken up by this rough handling.

Our first four stories and chapters have been comparatively abstract and conceptual, although they usually have had concrete referents (the factory, the clockworks, the family, the stage). Here, the story is deliberately instrumental and graphic and haptic. Handles and probes and tools are much like the everyday instruments of the carpenter and plumber and gardener, of the surgeon and the physician, of the warrior and the artist, of the cook and dressmaker and tailor, and of the industrial engineer. Philosophers have been fond of somewhat more intellectualist accounts of observation, knowledge, and theorizing. But, for the practicing physicist, there is perhaps much greater allegiance to, and alliance with, the guilds and crafts. Physics is everyday work, done between breakfast and dinner, work requiring an apprenticeship before graduation into journeyman status and, along the way, production of a masterpiece such as a dissertation if one is to gain authority within the guild.[5]

What must the world be like so that we can get hold of it in terms of handles, probes, and tools – namely, through a craft of knowledge?[6] Crucially for a physicist, the world must have localized features, handles we might take hold of. And the world must be able to be taken as objective and "out there," itself more or less unchanged by how we get at it, its perhaps complex dynamical insides being hidden by a comparatively less detailed, more stable, hard-to-penetrate yet in the end handleable surface.[7] Now there is an intimate connection between what Nature is like (chapters 1–4) and how we take hold of it (this chapter). Specific handles (as degrees of freedom), probes, and tools go along with notions of a factory, or clockworks, or kinship system, or theatrical stage.[8] Put differently, built into each model of the way Nature must go are handles for taking hold of Nature. And built into each notion of a probe is a notion of what it could possibly get hold of. Of course, this congruence between handles and Nature – an instrumental analogy – has required great artifice, the development of experimental and theoretical devices that mirror each other. For example, particle accelerators probe the world by shooting particles at a target or at oncoming other particles; we then note the patterns of scattering. And modern theoretical particle physics may be seen as a justification of taking the world as simply a matter of scattering processes (and hence those Feynman graphs [see Figure 3.1], which look like scatterings). So technology and grammar and theory are intimately interwoven.[9]

In part by recasting the material of earlier chapters, I want to describe a variety of handles physicists employ – the right degrees of freedom – and then the nature of probing and of tool and craft, and then how physics justifies itself in these terms. Surely lots is left out by the restriction to probe-ability; but, as we shall recurrently suggest, that lots is just what is not physics (but chemistry or engineering or . . .). Again, the description I shall provide, that of a complex technical endeavor assimilated to everyday pictures of the world, is an of-course to the practicing physicist.[10] It is just the way things are or almost are.

SOME HANDLES ONTO THE WORLD

A handle is a way of taking hold of some thing. If you pull on the handle, that thing comes along more or less well; and by tapping or shaking the handle, that thing may begin to vibrate. Of course, you need to set up the thing so that it is handleable; namely, it has to hold together well enough so that when you pull on the handle the thing does not fall apart, nor will the handle come off in your hand and become disconnected from the thing. Imagine tying together an overfilled suitcase so that it does not fall apart when you pick it up by its handle. Or, imagine rigging up a tuner control in an old-time radio – making sure the linkage is reliable and is connected to some critical internal elements so that when you turn a knob the station changes reliably and reproducibly.

I want to review some handles appropriate for a particle (namely, particle properties as handles), a crystal (stocks and flows), a gas (response functions), and the moment of transition between phases of matter (scaling parameters). Probes, it should be noted, are designed to take hold of those parenthesized handles. Along the way, I want to say more about what it means to be a right degree of freedom, namely, the specific traditional and conventional handle-sets; what it means to take the world as it is handled, namely, physics; and, how observation and knowledge may be conceived of as a matter of handling, namely, experimentation. Put simply, the right degrees of freedom allow us to take the world as if it has handles, and they lead to simple observation and orderly knowledge.

Particles. Handles onto a particle include its mass and momentum (which then specify its energy), and perhaps its charge, angular momentum, and other quantum numbers such as charm or strangeness. Even black holes might be treated as particles, ones which have mass, angular momentum, and charge.[11] Each handle tells you how the particle will interact with a probe, so that a particle's electrical charge "couples" (gets attached one to one, like railroad cars) to probing electric fields, and its strangeness couples to the weak interaction field. If a probe can take hold of a particle's handles, then it can literally move or change that particle. So an electric field exerts a force on a charged particle, and the particle's acceleration is proportional to its charge.

The goodness of a handle may depend on the environment or the vacuum. The handles onto an electron in otherwise empty space or in a single atom would seem to be inappropriate when it is embedded within a metallic crystal. For the embedded electron is so involved with all the other electrons and the atomic nuclei of the crystal, it would seem to have no well-defined handles of its own, there being no way of taking hold of it and it alone. However, it turns out to be possible to define a particle-like object within that crystal (a *quasi*particle) that behaves just like an electron, with handles like mass and charge. The quasi-electron is said to consist of an ordinary electron with a cloud of charged particles around it, and so the effective mass of this cloud is different from that of a bare electron. And, otherwise, it behaves "just like" an electron.[12]

Crystals. Handles onto a crystal lattice include its modes of vibration and their tones or frequencies, as well as the crystal's temperature. Those vibrational modes (like the shapes of vibration of a drumhead) are in part determined by the generic orderliness or symmetries of the crystal lattice, and in part by the details of its component atoms and molecules. Again, each mode of vibration is a handle that can couple to the outside world through a resonant probe ringing at that mode's frequency of vibration. Say we kept an account book of those handles, an account of which modes of vibration are excited and by how much, the score representing the notes and their loudness – a musical inventory of sorts. These accounts are called occupation numbers, designating which modes or quantum states are excited or occupied and by how much; they are analogous to stock-and-flow measures for an inventory of goods in a warehouse.

Now if the crystal were in thermal equilibrium, then its temperature handle, *T*, turns out to completely determine those occupation numbers. (Technically, letting *Energy* ≈ frequency of vibration, then the probability of excitation or the average occupation number is in general a simple function of $e^{-E/kT}$, where *k* is Boltzmann's constant.) So, if our probe were thermal rather than resonant, and if it were to just heat up the crystal, the response of a changed temperature fully determines the changes in those occupation numbers (and the energy required to heat up the crystal is the sum of the increased excitation of each mode).[13] Coupling to the outside world is through that very general thermal equilibrium, rather than through the very specific resonant handles for each mode. And so many seemingly independent handles are reduced to only one, the temperature. The music of the spheres is thermal, so to speak. Of course, we are still welcome to just hit the crystal, a non-equilibrium coupling or handshake, but the response is likely to be too complex to be readily interpreted. For the warehouse, the thermal case would be as if there were a normal mix of product demand, the mix's details dependent on the general level of demand. And so all one would need know is that general level of demand. We can, correspondingly, have a run on a particular product line, disturbing our normal product mix.

The particle physicists' physical vacuum is often treated as like a crystal lattice, or at least as a space pervaded by order (the theme of chapter 4). Handles onto such a crystal are the occupation numbers and the probing-induced changes in those occupation numbers, both of its various modes of lattice vibration and of the electron-like objects within that lattice – namely, the demographics of the various particles and quasiparticles in this vacuum (which are called, respectively, collective modes and elementary excitations). Each particle is a representative of the vacuum's orderliness, the particle's equivalent vibration frequency being proportional to its energy. And, again, there are cases when this populated vacuum is given just a temperature (as in the Big Bang scenario of the early universe), a temperature which then determines the occupation numbers of all the varieties of particles.

The number of independent handles we may attach to something depends on matters of both setup and convenience.[14] As we have seen, if there is an equilibrium defined by a temperature, then many of the once free handles are more or less determined. The occupation numbers are no longer free but are prescribed by the exponential rule. (Fluctuations do occur. In general they are small, effecting only minor excursions around the equilibrium occupation numbers. But we know only the spectrum of those fluctuations, not about each one individually.) Similarly, if a gas is in thermal equilibrium, its pressure and volume now depend on each other (Pressure times Volume is proportional to Temperature), and so they are not both free. In each case, we have in effect tamed some if not many degrees of freedom, for they are no longer independent handles.[15]

Gasses. Handles onto a gas may include its pressure, temperature, volume, and composition. Volume is measured with rulers, while pressure and temperature are measured with gauges and thermometers. But in terms of probing, pressure, for example, is measured by how hard it is, how much work is required, to change the volume – that is what a gauge measures. Probing – here think of that fork being pushed into a potato – is a matter of how hard it is to change something.

Handles on the expanding spacetime universe – if it is considered as a "slowly" expanding gas – are its temperature, density, and expansion (Hubble) velocity. As we have discussed, a good handle onto a red-hot piece of material is just its temperature; the material may be considered to be a gas of electromagnetic radiation (light or photons or oscillators) at a certain temperature, independent of the details of its chemical composition. If a gas is flowing or turbulent and not uniform, then many of its obvious handles (velocity, density) are field-like, in that they have different values at different points. But these handles are not all free. As we have seen in chapters 1 and 2, there are rules concerning continuity and flow (in general, what comes in must go out) that connect the handles at one point with those at another adjacent point.[16]

Analogy. Physicists readily entertain a powerful principle of analogy, one that says that if the handles onto something are like the handles onto a gas or a crystal or whatever, then we may think of that something *as being* a gas or a crystal or whatever.[17] This is a generalization of the analogy of substance we discussed in chapter 4. Recall that an electron inside a metallic crystal is called a quasiparticle; it has particle-like handles much like those of an electron in empty space. Or, the vibrations of a crystal can also have particle-like handles – we treat them as sound particles or "phonons" residing, so to speak, in that space defined by the crystal lattice. And as we might expect, such a vibration particle may then be shown to be able to bounce billiard-ball-like off another particle, whether that other particle be another phonon or an electron or a photon, for example.

Moreover, physicists do not much worry about what something is "really" like inside itself, what it ultimately is. They surely would like to be able to probe much more insistently and finely than they just now do, so revealing currently hidden or walled-off degrees of freedom. But for now, they get hold of what they can and plan to do better experiments as they learn how. And, in any case, it is how something may be taken hold of that tells them what it is.

For example, for the physicist, one of the most influential models of a handle or a degree of freedom is provided by the spring, since very many situations may be analogized to a spring's vibratory behavior – and so these situations are taken as physically interesting in just this way. A spring – a harmonic oscillator, to use the term of art – is gotten hold of through two degrees of freedom: the frequency of its vibration, namely its ringing tone or harmonic; and its amplitude of vibration, in effect how loud that tone is. Recall that the modes of excitation of a vacuum can also be known by the energies of the particles that may be present in the vacuum (or their frequency of vibration) and by the occupation numbers (measures of loudness).[18] So, as indicated earlier, we might come to think of the world as being literally demographic, populated now by little springs or oscillators, the dynamics being a story of stocks and flows, much as in accounting.

Put differently, when things are in equilibrium or set up to be an empty stage or a ground state, or almost so, small disturbances often lead to springlike oscillations around that equilibrium point. So physicists explicitly justify why harmonic or springlike handles are often a way of getting a good handle onto a physical system. Springlike handles are said to be "linear," meaning that small disturbances or changes have small and proportional effects – as for springs on a scale or as represented in a straight-line graph – just what we might expect at equilibrium or from a ground state (one of those empty stages or vacua). Now, a good experimental setup may be thought of as objective, out there, more or less independent of us.[19] Such springlike handles allow us to readily extrapolate to zero influence and, hence, objectivity.

Of course, analogy is limited. Just because the handles onto several objects are the same does not mean that things will turn out to be the same "inside"; that is, when we do eventually look more closely. As we know more about two similarly handled objects, subtle differences in their handles both in quality and quantity will develop: nondegeneracy shows its hand once more. So one wants to get inside, eventually, to find out what is really going on – presumably, of course, to encounter only other handles and analogies. Eventually, if not now then later, we won't be able to get further inside, at least for the moment. The differences of internal mechanism or workings or features are again of no consequence. For the moment, we do not see any evidence of that difference. To boot, physicists are tempted to push any analogy as far as they can, since once it proves productive they are naturally reluctant to give it up. They set up experiments so that they do not see any gross evidence of that difference between the physical situation they are studying and some other situation or a model or analogy. That is what they usually mean by a good setup. A sophisticated physicist has a reasonably good sense when such setting up is illuminating rather than hiding what matters: when an effect is noise to be suppressed so as to maintain the analogy rather than a phenomenon to be highlighted, showing us how to go further inside. But not always.

Phase Transitions. Finally, recalling our discussion in the preface, let us consider handles onto the setting in of the permanent magnetization of a piece of iron as it is cooled down from so high a temperature (say, above 1100 °K) where it cannot retain such a magnetization. Presumably, the handles include temperature and degree of order (magnetization). But these are not, it turns out, the most perspicuous of handles when we are very close to the transition or critical temperature (T_c = 1043 °K). It turns out that we then want both a temperature handle that is sensitive to the distance to the transition temperature and an orderliness handle that allows for magnetization that occurs at spatial scales smaller than the whole piece of iron.

Technically, temperature must be modified, becoming a "reduced" temperature, $t = (T - T_c)/T_c$ – to emphasize that it is the distance to a critical temperature (T_c) that matters. And magnetization is articulated, made more subtle as a notion, by a measure of spatial influence: the probability that an atom at a certain distance from a given atom will have its magnetic spin lined up with the given atom's spin, so contributing to that overall magnetization. (Recalling our discussion in chapter 4, the crucial distance is called a correlation length, *l*, designating the distance beyond which the probability of correlation, and so the possibility of overall coherent magnetization, begins to fall substantially.) There are then additional handles, called scaling parameters, say expressing the strength of influence in terms of the size of the correlation length (stronger influence meaning a larger *l*). Technically, in terms of *t* (and not *T*!), $l \approx |t|^{-u}$, where the scaling parameter or exponent *u* is typically a number like 4/3.[20] At $t = 0$ ($T = T_c$), *l* is infinite, since now it is possible to have magnetization in bulk, namely throughout the iron bar (which is "infinitely" large compared to an atomic scale).

That these modified handles were the right handles, the right way of getting hold of a bar of iron about to be magnetizable, the right degrees of freedom, was discovered in experimental studies of phase transitions where they proved to be good handles. Those experiments found what is called universal behavior – many different transitions looked the same when expressed in terms of reduced temperatures and scaling parameters. Eventually, those became the most natural of handles, used to express much of all we know about such transitions, conceptually and theoretically as well as experimentally. And so such handles become obvious, part of a new conventional tradition. In earlier work there were somewhat more awkward and less versatile handles, natural enough until new ones came along which showed how artificial (so to speak) the old ones were.

Knowledge Is Handling. I have argued that physicists try to set up situations so that they are handleable, namely, described by good handles that are the right degrees of freedom. It is a holding-on-to that nicely couples an object to the outside world, and allows us to shake the object, usually gently, and in this manner to learn about it – what is called probing. And we might hope that such handles are sufficiently distinctive (nondegenerate) and more or less free and independent of each other, so that we know just what we are shaking or probing (Maxwell again).

Moreover, there is a very great and productive temptation to analogize: to take anything that can be handled in a certain way to be just like anything else that can be so handled. Here, knowledge is unapologetically analogical and generic, those generic ways such as the harmonic oscillator model defining what we mean by physical objects, namely, objects suited to physicists' modes of handling.[21] Still, such knowledge is, in the end, a matter of specific detail, just which analogy applies and how it is fulfilled. When we set up a system so that it is handleable, suitably arranging the experimental conditions or theoretical constructs, we are modeling the system by one of the ways (perhaps of the first four chapters) that the world must go. But just how it ends up going is open to inquiry. If things go differently, the consequences and meanings can be quite different. Knowing the repertoire of analogies and kinds of handles is never enough. It is only through experiment, going out and getting a good handle onto the world, that we find out about it.

Put differently, the actual doing of physics is both a ritualized practice and a compromise with the actual world. If we follow appropriate practices or rituals, we might learn about the world; it might well make sense under the rubric of physics. But we have to actually follow through, do that ritual work; for abstractly the practices do not tell us just how the world goes. And it is the "just how" that in the end matters. And in doing that ritual work, we teach ourselves to make a compromise with the world, setting up conditions and constraints so the world might be physical, namely, that it be a form of Nature.

In sum, if we know only what we can get a handle onto, and if handles are of particular sorts, then knowledge is a matter of entering into a hard-won covenant, so to speak, with one of the archetypal representations, models, analogies, or ways the world must go. (To be Kantian, knowledge when we achieve it is categorial and transcendental.) That covenant is both conceptual and experimental. Physicists know they are "on" to something when the physical system as set up really does go or can be seen to go in one of these archetypal ways, or perhaps not. Moreover, they observe the world by probing it, by shaking those handles and seeing what happens in response. There is nothing else physicists could take hold of by probing except those handles. Now, we need to further examine just how probing takes place.

II

> [Going at the nucleus with tongs vs. hammer . . .] We think of nuclei as made up of nucleons [protons and neutrons]. . . . If we want to investigate vibrations associated with spatial coordinates [of the nucleus and its component nucleons], we should, figuratively, hit the nucleus with a hammer and listen to it ring. If we want to investigate the vibrations associated with a change in state of a constituent nucleon, we should flip the nucleon with tiny tongs and listen to the nucleus ring. The *(p,n)* reaction at 100–200 Mev is our tongs. It excites giant Gamow-Teller resonances . . .[22]

PROBES

To preview what will come: A good probe requires a good vacuum, a vacuum that responds gently to small stimuli, a vacuum that is swept up well enough so that irrelevant degrees of freedom, in their providing additional and arbitrary responses, do not get in the way of our discerning a well-defined, gentle response to the probe: namely, the vacuum is marginal, stable, and nondegenerate. Then a probe will be able to take hold of a handle, yet not disturb the orderliness of the vacuum too much. Such well-designed probes are strong enough to be informative, but ultimately weak enough so we might see the world and conceive of it as absent of their presence – so we might think that in probing we are merely observing the world out there. That is how there can be objectivity. Now, probing, observation, and objectivity are actual practices, things physicists learn to do and to do well. They are craft skills. Put differently, the meaning of these terms is defined conventionally, by those practices that actually work to provide physicists with a Nature that can be told about in the kind of stories I have discussed in earlier chapters. Of course, Nature itself is still surprising enough to force physicists to face the truth – that those stories are never the whole story, at least as we first understand them.

Probing pricks the world and so populates it with recognizable phenomena and objects – or so physicists learn to think of probing in this way. One probes something by shaking or pulling on one of its handles, and then looking for a response or a resistance. As I have indicated, a response is by convention something recognizable and simple, apparently directly connected to the original probe. Probes actually probe by supplying the world with just enough energy (as well as angular momentum, charge . . .) in just the right way to take hold of a specific handle – a degree of freedom – and shake it or excite it: to accelerate a particle, to change the temperature or pressure, to stimulate a fluctuation, to change an occupation number or create a particle (and hence my speaking of "objects" above), or to ping or tap a spring so it vibrates and we hear its tone.[23]

In general, probing changes things marginally, just a little. And the probe is selective, not supplying much energy to, or exciting or pinging, most of the handles; and so we do not see responses characteristic of them. Most of the degrees of freedom remain tamed and hidden. If we think of a probe as literally part of an experimental setup, we might say that the setup holds down (or clamps down upon) those other degrees of freedom; they are not allowed to change and so they remain hidden. In effect, to probe the world is to uncover and so create recognizable responses, ones that are regular and reproducible. A good probe gets at just one degree of freedom, and it does so in a gentle enough fashion that we may conceive of how we might "extrapolate down" to zero interaction and effect, and hence to a notion of objectivity, the probe in effect not being present (and we say this, in effect, even in quantum mechanical situations, for which see the note).[24]

Physics taken as probing is a cognitive science, providing for literally poignant or haptic knowledge of the world, namely, a way of getting hold of or of touching particular features which have significance for a whole system. Hence, probes and the handles that go along with them are indicators. A good handle is sufficiently sensitively attached so that lots of information flows from probing it; yet, paradoxically, often the system remains more or less as if untouched, if the probe is gentle enough, or we may infer what the system was like before we probed it. Correspondingly, Nature will seem objective when it is probed if the handles are attached marginally, stably, and nondegenerately. Responses are directly related to, actually proportional to, probings (the responses are small if probings are small); the system does not readily turn into a new, very different configuration when it is probed; and, responses are reproducible.

This description of probing is drawn from several of the usually mentioned conventional examples of physical probes: adding heat or doing work on a system; ballistically shooting charged particles into matter; and pinging a tuning fork (namely, harmonic excitations). Such examples model for physicists what they mean by probing the world.[25] Yet some physical probes are, as we shall see, explosive and inelastic, and the world does not just bounce back or extrapolate down to zero from their influence. To retain objectivity, physicists model just how such probes are disruptive, and in that way they hope to be able to conceive of what things were like before the explosion. (Think of the problem that is to be solved by a forensic bomb squad.)

These examples recall the examples of handles in part I, but now from a different point of view – and for good reason. Probe and handle are reciprocal roles. Note, for example, that a charge can be either a handle or a probe. If the charge is a handle, the probe might carry a field which can interact with that charge. If the handle is a field, then the probe might possess a charge that couples to that field.

If a probe gets a handle onto an object, that handle in effect reciprocally probes the world around it. Physicists would say that the interaction of probe and handle is linear and multiplicative (Probe × Handle), allowing both for interchange of the two roles (Handle × Probe) and for going down to zero probing and so zero interaction. Of course, in actual practice probes designate what *we* are using to find out about the world and its handles, and so in practice probes and handles can be very different roles.

When physicists probe a gas or solid by heating it up (adding heat to it), they look for a response in a changed temperature, pressure, or composition, or a change in the orderliness of the system (magnetization, for example). Now, as we have said, in general such small probes have small effects in response – at least if there is equilibrium.[26] When they do not; when the responses are very big, in effect infinite when compared to the probe; when probing seems to be nonmarginal: physicists say that there has been transition in phase (melting, for example) or, in other kinds of probings, that a threshold has been exceeded or a resonance has been hit or the equilibrium is unstable, or an inelastic interaction has taken place. Something much more, something structural, must be going on. So probes are divided into those that are marginal and linear and those that are discontinuous or resonant or transformational or inelastic in their effects.[27] As might be expected, very substantial effort is put into understanding the discontinuous or inelastic cases.[28] The aim is to classify them – for example, phase transitions of the "first" and "second" kinds – and to describe generic patterns of their appearance. So what might at first seem like an extraordinary consequence of probing now becomes a patterned and parametrizable response, of a particular sort.

If a probe scatters off an object, and so probes it, and that object remains much the same as before, a simple story is sufficient. But if the object would seem to be transformed, as in a chemical reaction, we'll need to incorporate a notion of interaction for which it may not be possible to extrapolate down to zero interaction.

A second kind of probe is called a test particle. Again, it is supposed to be quite small in its effect; its presence should not affect the world much. Rather, the test particle should display the presence of a field and be literally moved by it. We say, the particle "feels" a force due to that field; it couples to the field. Charged particles, such as electrons, measure electric fields; moving charges or electric currents or magnets measure magnetic fields; pressure gauges measure hydrodynamic flows. A good test particle probes by being reliably affected by the field, and so it has to be rather uncomplicated itself (so it does not change on its own, so to speak).[29]

A third kind of probe is the hammer and tongs. Were something made up of springs or tuning forks, or suspected of or taken to being so composed, then if we hit it we will excite those springs. And so we hear in response the collective sounds of their ringing. We hit a drum, and what we then hear is the ringing of some of its modes of vibration. We can literally hear some features of the shape of a drum, such as its area and the length of its perimeter.[30] Now if we can pull on just one spring, or excite just one mode of vibration, one pure tone, we can determine both its size of oscillation (its loudness) and its tone or resonant frequency, even if other springs are set into overtone vibrations. And size and tone are the right degrees of freedom here.

Technically, physicists *observe,* that is, they actually see the consequences of these probings through pointers moving, light flashes on a screen, and bubble chamber tracks, namely, through electromagnetic interactions, however the probe itself does its probing. Physicists' observation – the grammar of observation – takes advantage of several features of electromagnetic interactions: the possibility of rigid structures (that is, laboratory apparatus and equipment); the smallness of the electron's charge (so a charged object can be a gentle probe or test particle itself, or it can be a sensitive nondisturbing handle onto a probe, showing what that probe is doing, as in a bubble chamber track of a probing particle).[31] Moreover, we want robust data, information from the probe (or an object) that is independent of the detailed history of how we got that information. What I have been suggesting, and here it is just a suggestion, is that probing and observation, however they may be thought about by philosophers, are physical processes. And if the electromagnetic interaction were very different than it is, the grammar of probing and observation, roughly in accord with our everyday usages, might not be applicable.

Whether probes are thermal, particulate, or harmonic, what makes their actions probings is that they are small in size and yield distinctive, mostly marginal responses. Again, probing usually gets hold of a very few degrees of freedom, preferably just one; the rest of the degrees of freedom are in effect tied down. Typically, all of this work is done by properly setting up one's experiment (and theoretical structure) so that it provides a good vacuum or ground state or equilibrium in which the probe's influence is specific – since the probe is now more or less in accord with the vacuum's orderliness, and its influence can only be felt by what is left on stage (Somethings). Moreover, if physicists' archetypal models of such probing include gently heating up an ideal gas, placing an electron in an electric field, and hitting a tuning fork, their theories of thermodynamics, electromagnetism, and mechanics say why these probes will tell us about the world, why they get at the right degrees of freedom. So, for example, electromagnetic theory prescribes the force a charged particle will feel in an electric field: Field-Strength-at-that-point times Charge, where the handle onto the electric field is that field strength at each point. The theory justifies such a probe by saying why a test particle will not disturb the field too much and why it feels the field at just a point. And by its formulation, Maxwellian electromagnetism is in terms of those electric and magnetic fields: They are the right degrees of freedom.

OBJECTIVITY AND INELASTICITY

When physicists speak of probing a physical system, they expect that the system is not too much affected by their probing, their probe being of certain canonical sorts. That is what makes for a good physical system, one that is properly set up. That gentleness of effect goes along with our intuitive sense of what a probe ought to be like. And, correspondingly, the world is objective, out there, to be observed by us. Now if the probe is not marginal but rather discontinuous in its effect, physicists still believe that given a suitable experimental setup they can in the end conceptualize the world as if no probe were present – as if they were not present.[32]

Again, of course, actual probes are never so silent and gentle. So if the probe is marginal in its effect, physicists believe they may extrapolate down, roughly proportionately, to zero interaction. If the probe is discontinuous, they use models that show how a system might respond discontinuously. If the situation is such that quantum effects are dominant, then probing *must* change things (namely, the Heisenberg principle that an informative probe or measurement must have an unavoidable nonzero effect on a handle). But quantum mechanics then prescribes how we are to go from information on a probed situation to the original untouched situation or wave function or state vector – if we have some idea of its generic form.[33] And in general relativity, the space and time in which the probing is done is determined by the handle itself, the "metric tensor" being the handle onto and a function of gravity. (Technically, the metric tensor *is* gravity.) In effect, that metric tensor tells physicists how to use small test particles as probes much as they are used in the more conventionally objective flat space we know every day.

Now if a probe is sufficiently energetic and specific, it penetrates shields and so excites and liberates degrees of freedom, such as particles that are ordinarily held in and tamed. We might treat those produced objects as if they were "already there" before the probing released them, as we have discussed already. But how we might conceive of their being already-there requires an explicit model of how they once fit inside, as for example how electrons are bound to a nucleus in an atom, when the electrons are kicked out by a photon, or again a model of chemical reactions or of the production and annihilation and exchange of elementary particles.[34] (Again, imagine looking at the ejecta from an explosion and inferring the original configuration.)

Now in general, such an energetic probe may not leave things even close to being the same as they were before, and it is said to be highly inelastic and irreversible. The system could not just bounce back to its original state, for the probing spews forth a great deal of material that is not readily restorable to its original place. For, although the probe might even work along just a few degrees of freedom, poignantly stimulating the vacuum, the system falls apart and the response typically shows itself through many degrees of freedom. Hence, reversing such an effluvium, getting all the horses back into the barn, is quite unlikely.[35]

To allow themselves to believe they are still probing, physicists develop ways of systematically or canonically dealing with inelasticity and irreversibility. One way is to study the zoo of objects that are produced, reveling in its variety and richness, both abstractly and in detail. (But too much detail would be considered "botany," here meant pejoratively. So they may study that zoo's population statistically.) A second way is to radically tame that variety: say, by treating the probed object as having been heated up by the probe, and hence it is a fireball (at a given temperature) that then evaporates off that zoo of objects; or, by grouping the zoo of objects into distinct collections (such as explosive but coherent jets of particles emanating from the source); or, by paying attention to only one of the produced objects, ignoring the rest.[36] In each case the probe now does just one thing: it heats up the probed object, or it produces just two opposing jets, or it takes hold of just one handle (by averaging or summing over the rest). And so, sometime, the qualitative changes might even be extrapolated back to zero effect.

A very great deal of conceptual and theoretical effort goes into figuring out how inelasticity and irreversibility may be "modeled away": physicists develop models of the original system in which they can figure out the effects of inelasticity. So if there is an explosion, how are the properties of the ejecta related to the original object's properties? The idea, always, is to be able to probe the world yet not affect it, at least in the end, or at least in principle if not in practice – even if we have convinced ourselves that some probes in actuality must be unavoidably disturbing and destructive.

In sum, probing makes for an objective world, a world we may eventually consider as out there, independent of us, more or less unaffected, at least qualitatively, by our actual probings. Experimental probes and detectors may well produce the phenomena physicists call Nature by their kicking the world and displaying its responses. But, in the end those probes and detectors must be forgotten, the phenomenon (the physical world) now becoming a fetish, alienated from its mode of production.[37]

PROBES AND HANDLES

Probes take hold of handles and pull or shake them a bit. A good probe gets hold of one handle; and if the handle is well chosen, the response will be recognizable as a response, and be simple as well. Once physicists have probed an object and found out about its handles, they then take those handles as being intrinsic to the object (although a handle is only seen when we are probing), and so our knowledge is fetishized and thus objective. The handles are out there, independent of us. What something is, and how it interacts with the world, are the same; or, what something is, is literally denominated by its interactions. This commitment to handles as such is just what Marx meant by commodity fetishism in the economy; what Bishop Paley (1802) more or less assumed (and Hume had earlier mocked) in saying that the complexity and orderliness of the universe as we know it is a sign of God's design; what an anthropologist presumes by describing a society in terms of the exchange of women; and what we do when we take what is on stage in the theater as all of what counts.[38] Of course, everyone knows there is something more besides handles: either stuff that is too complicated, at least for now, to be seen as physical (or economic, mechanical, societal, or narrative and theatrical) or stuff that is beyond such knowledge. Still, physicists know how to do lots of work with the handles they do command, and that work is productive and interesting.[39] How might we describe the work they do? Physicists like to think of that work as a matter of employing a toolkit, their being engaged in a craft of science.

III

TOOLS AND TOOLKITS[40]

> Those of us who were fortunate enough to watch Enrico Fermi [1901–1954] at work marveled at the speed and ease with which he could produce a solution to almost any problem in physics brought to his attention. When he had heard enough to know what the problem was, he proceeded to the blackboard and let the solution flow out of his chalk. He kept in trim by doing a lot of problems, either for the courses he taught, the talks he gave, or the papers he wrote. Most frequently, he worked out his own solutions to problems he heard about, in seminars, or in discussion with those who came to talk physics with him. Fermi's solutions were almost always simpler and easier to understand than the ones obtained by the person who raised the question in the first place.[41]

Physicists actually think of themselves as employing a toolkit when they probe the world, take hold of its handles, and make sense of what they find out. They readily speak of their kit of analytic and modeling tools, and the description above is of a master craftsperson by one of his students.

Of course, any toolkit comes along with a practice of tool-using – how to do the work and how the work is organized, and a sense of what it is the craftsperson is up to. Surely a carpenter's kit will have hammers and chisels; but carpenters use nails and work with wood; they know how to swing a hammer; they work during regular hours along with other crafts; and they are in the business of constructing something. Tools are both objects, as are hammers, and instruments and functions, as when a rock is used as a hammer. The term "toolkit" suggests a small number of versatile tools meant to work together.

It is within a context of tool and craft that a practice, such as doing physics, is clear, natural, and simple. Journeymen take on jobs amenable to their skills and formulate their projects so that they can work on them using their kit of tools. So physicists will persist in setting up stable equilibrium situations; they can then probe the world gently and get small or linear responses, and most of the degrees of freedom remain hidden. When they ballistically shoot particles at other particles, they usually aim to have one particle hit just one other particle. The models provided in each of the first four chapters define sets of tools and modes of handling and probing, so we may in each case construct a physical world. Physicists then justify their capacity to encompass Nature by giving an account of why there are divisions of labor, partitionings, family structures, and hierarchies of forces.

A PHYSICIST'S TOOLKIT

A physicist's toolkit might be divided into three parts: mathematical or calculating, diagrammatic or picturing, and rhetorical or describing. The trivium of mathematics, diagrammatic, and rhetoric correspond roughly to structure and number, pattern and image, and argument and language. These categories represent a long-standing tradition within rhetoric that saw logical and pictorial knowledge as being in opposition, and rhetoric itself as being the art of address (here, the art of addressing Nature).[42] This particular kit is perhaps best suited to the paper-and-pencil conceptualizing, explanatory, and problem-solving work of a physicist.

Mathematics is perhaps most apparent as a skill and a set of tools. Some of the deepest principles of Nature, as physicists understand it, are built into the subset of mathematics that can do this practical work of counting degrees of freedom, classifying them, and accounting for how probes gently excite them.

Diagrammatic or picturing tools are figures that represent Nature, with which we may "write down" and then "read off" the physics (as a physicist might say) – once we are well enough trained to be graphically literate. These figures are either geometric and spatial (as in Figure 3.1), or arrows indicating the forces on an object, or algebraic and symbolically patterned ($\partial_\mu \partial_\nu F^{\mu\nu} = 0$), and often they are in effect modes of representing symmetries and the conservation of flow.

If rhetoric is the art of address, here the rhetorical or descriptive tools address Nature in terms of that factory, clockworks, kinship structure, and theater; but the actual tools are rather more generic media, objects, and interactions. And it is in this generic way that Nature becomes something physicists can talk about.

When physicists approach a situation as physicists they immediately take it *as* a physical system or a medium, such as an orderly crystal. As we saw in chapter 4, the media are often treated as sequences tending to greater symmetry and higher temperature (for example, solid/liquid/gas or atom/nucleus/quark). As the temperature rises each medium "melts" into the next one, and so unfreezes some degrees of freedom.

While settling on the medium, a physicist immediately identifies the things *in* that medium – excitations or Somethings, such as particles and fields. A physicist calls these things objects; and what seems crucial about objects is that each object is localizable in some space, and it is often a demographically summable individual, even if it, itself, is made up of much more complicated not-so-arithmetically-additive stuff. (Even if the object is a field, it usually possesses a form of locality and of additivity: A field's changes at one point depend on what is happening nearby; and to combine the effects of two fields, one adds field values at each point, what is called linear superposition. The attentive reader will recall that in chapter 1 we allowed only particles to be objects; here we have generalized, reflecting actual usage).[43] A medium can be taken as an object in another medium, so a gas of electrons is said to fill a metallic crystal (and we may speak of the temperature, pressure, and so forth, of that "Fermi gas"). And there are modes of transformation among the objects: A particle can come to be seen as a wave or a field; or a wave becomes an oscillator; or a field becomes a population of particles. So a crystal can be seen as being a vacuum populated by particles such as phonons. Because those transformations are not only at a single scale (technically, metaphor), but between scales (synecdoche), an atomic hypothesis tells us to look more closely.

In actual usage, tools are taken quite specifically, then to be modified or generalized for the project at hand. A crystalline solid might be salt (NaCl), an atom will be the hydrogen atom, a gas will be an ideal gas, a field will be Maxwell's electromagnetic one, and an interaction may be inverse square.

Media and objects interact, perhaps eventually being transformed into something other than what they were initially. Interaction requires actors such as particles, means such as forces (which may well be transmitted by other particles through exchanges, as in chapter 3), and modes of resolution, on a suitably empty stage. So, eventually, interaction may be seen to have expended itself and be resolved. The now-interacted and perhaps transformed media and objects may once more go along independently – until the next episode of interaction.[44]

There are a number of strategies of address: find a good vacuum or ground state, such as a crystal lattice; or look for a good division of labor, or modular partitioning, or family structure. We have often mentioned two other strategies. The first is that of analogy and heuristic. If you can get away with treating a blackbody as a gas, or light as a particle, or nuclear forces as simple central forces, do it and believe it. A second strategy concerns the plenitude of nature: Everything that can happen will happen. Here is the physicist Tullio Regge speaking to the writer Primo Levi:

> I, however, would like the universe to contain or eventually contain in its history, which is infinite, every possible object that can be conceived of by the field equation. A statue of Primo Levi made of Tibetan olive oil cooled to less than two hundred degrees certainly must exist in some part of the universe: it is an extremely improbable object, and it will be necessary to travel a certain number of billions of years in order to find it, but it exists somewhere. The universe is infinite because it must allow for everything that is permitted, because everything that is permitted is obligatory.[45]

And if something does not happen, there must be a good reason why – so look for that reason.

Put differently: Get hold of something that will mostly stay put, and find the natural units or kinds of things within it. Use all you know about one situation to investigate another that seems to be like it. Assume that the world is tractable, one way or another, to your understanding. And, finally, whatever does not happen is actively forbidden.

The structure of this toolkit is actually quite conventional. It is drawn from descriptions of classical rhetoric. Media correspond to genres, objects to tropes, interaction to plot, strategies to modes of address, and standard methods to commonplaces. Scientific explanation is rhetorical, in just these terms. Moreover, what is distinctive about this rhetoric or mode of description is that the various media, objects, and interactions may be combined in seemingly arbitrary ways – *x* medium with *y* object with *z* interaction mechanism. Such modularity may be taken as a sign of the alienated and fetishized world of modern science, its alienation and fetishism signs of a wondrous human invention rather than signs of decadence and sacrilege.

As might be expected from its design, this toolkit slights some crucial tools such as actual laboratory equipment, in part because it is concerned with paper-and-pencil problem-solving activities. For experiment and empirical investigation, there must be additions to the kit: namely, tools of inquiry and of recognition. For inquiry, there are logical tools, such as filters, for picking out prospectively interesting events; amplifiers for making them more prominent; and actual stimulators or probes. For recognition, there are tools for simulation, in order to figure out how, say, an electron might act in an apparatus; tools to manage dissimulation, dealing with how chance and fakery produce what appear as genuine events; and tools for emulation, to set up situations so that they are "the same," so as to provide for reproducibility (namely, good data). In a good toolkit, the theoretical and the experimental tools are intimately connected, so that a probe might well be a particle, and filters and amplifiers are almost surely interaction mechanisms.[46]

A story of craft – and a rhetoric of handles, probes, and tools – says that physicists get at the world by artfully taking hold of it. Rather than an epistemology based on sight and on passive objects, here the epistemology is based on touch and grasp, in effect working with your hands.[47] Of course, what is remarkable about this story is that important parts of Nature go along with it, that what we have informally taken to be Nature is probe-able, possesses handles, and may be known instrumentally.

SO FAR

I have been describing what Kant would have called the transcendental conditions, or Wittgenstein the grammar, for doing physics: the forms through which the work is done.[48] In effect, it is an anthropology, a description of a human culture. Yet, ideally, when practicing physicists read such a description, they will say: So what? Of course. That's right. Let's get down to real work. Even if they are thoughtful and reflective, their perspective is deliberately provisional and instrumental and unphilosophical – this is the way the world really is – as are most craftspersons' perspectives. The world must go this way, in part because it has done so before; and if it did not go in one of the ways it must go, we would not know, at least initially, how to talk about it as physicists. Whatever the world is really like, this is how we, as physicists, go about taking it; and that way of taking seems to be productive much of the time. It allows us to get at Nature. A tool or model or archetype lets us get at the world.[49] And then in our involvement in the world, the tool disappears; and so we are left with just the world, Nature as such. (We only discover the tool as such when it does not work or it breaks down.) And what we work with is not so much an abstract model or archetype as a concrete example, which of course is an example of that tool in its actual employment. If models are tools, then examples are tools in context.

Physicists also believe, although they do not feel obliged to deliver on all the details, that they can give an account of how the physical world allows the physicist to take hold of it. Such a congruence, between the world described by physicists and the possibility of their getting hold of it given their probes and sensitivities, is not a tautology.[50] If there is circularity, it is a rich and complicated circle. (Recall that "observation is electromagnetic" is not only about the smallness of the electron's charge, but also about the possibility of rigid apparatus, a not at all obvious consequence of electromagnetism and quantum mechanics.) More importantly, the experience of physicists is that whatever congruence there is, it is belied by the strain they experience between their conceptual and experimental capacities and the world itself. Nature resists our entreaties. There is work to do.

Objectivity, in this mode of working, is no casual fact. That physical forces allow for shielding and separation, that probing is a mode of observation, and that handles are a mode of knowledge mean in effect that an object is separable from us and can be taken as independent of how it is observed. Objectivity is practical and physical.[51]

Analogy, it seems, is destiny. My claim throughout is that these analogical ways, in their success, are testimony to ontologies of manufacture, mechanism, system and exchange, and theater, and now an ontology of tool and craft. Again we return to Adam Smith's factory and economy, with their division of labor, their varied individuals, their processes of haggling and exchange, their nicely defined assembly lines, and their industrial engineers (physicists!) rationalizing the whole process. In the next chapter we shall describe some of the ways physicists employ mathematical tools or machinery (supplely and inventively) to provide further analogies for doing physics. When physicists are up to one of these ways of taking the world, they are natives *and* they are philosophical anthropologists describing the culture of Nature.

SIX

Production Machinery: Mathematics for Analysis and Description

Philosophical Analysis and Phenomenological Description; Machinery and Production Processes; Naming and Modeling the World; Demonstrations and Proofs as Strategies of Explanation; Understanding "The Physics"; Analogy and Syzygy; The Mathematics and The Physics.

THE ARGUMENT IS: MATHEMATICS PROVIDES MACHINERY FOR modeling Nature, physicists customizing that mathematics so that it does the work of physics and of Nature, and along the way that machinery allows us to analyze and understand physical phenomena. So, for example, in mathematically modeling and so giving specific meaning to or "naming" freezing, or diffusion, or fluid flow, physicists discover better what they mean by those notions. In viewing Ising matter from many different mathematical points of view, physicists discover that Ising matter accommodates many modes of conception. The deep question then becomes how can Nature accommodate so many points of view (much as an everyday object accommodates many aspects and uses). Correspondingly, we are led to ask how are the different mathematical conceptions related to each other – what accommodates all of them. So there is an analogy between a physical analogy and a mathematical analogy, an analogy of analogies, what is called a syzygy.

> Philosophy [physical science as natural philosophy] is written in this all-encompassing book that is constantly open before our eyes, that is the universe; but it cannot be understood unless one first learns to understand the language and knows the characters in which it is written. It is written in mathematical

> language, and its characters are triangles, circles and other geometrical figures; without these it is humanly impossible to understand a word of it, and one wanders around pointlessly in a dark labyrinth.
>
> – GALILEO GALILEI, *Il Saggiatore* (The Assayer, 1623)[1]

While Galileo's claim would seem to describe much of modern physical science, in fact natural philosophy is written in many and varied mathematical dialects, some of which would appear to be different languages rather than dialects. The practical problem for the physicist is to discover at least one such language or dialect that actually works, and often this is a breakthrough achievement. In time, it is sometimes found that many such languages or dialects would seem to work, yet they are apparently very different from each other. The problem then becomes to discover the nature of the world that it will accommodate so many useful but very different modes of expression. What is the identity, the deeper physical or mathematical features of the world, that founds these manifold presentations or accounts?

PHILOSOPHICAL ANALYSIS AND PHENOMENOLOGICAL DESCRIPTION

The twentieth century saw the development of philosophical analysis and phenomenological description.[2] Philosophical analysis was an attempt to address traditional philosophical issues of knowledge or morality or beauty by becoming more precise about what we might mean by those notions. It used lots of examples, often from specialized practices such as particular natural sciences. By comparing closely similar examples, philosophical analysis was able to test out theories of what we might mean by "knowing something," for example. Initially, the aim was not so much to construct an overarching theory or system, but to make clearer some of the dilemmas and problems that philosophers had worried about over the centuries. Philosophical analysis heralded a new level of rigor and precision in philosophy, in which "local" theories replaced grand and global systems of thought.

Phenomenology is an attempt to describe our experience of the world. One of its deepest insights is that we know about the world *aspectivally,* through many instances, each different, and what we know about the world is not just those instances but a more general notion which allows for or accommodates each of those instances. Hence, we might know an apple by looking at it from various angles, cutting it open, eating it, watching it decay. And our notion of "apple" accommodates all those aspects yet is not reducible to them.

Mathematical machinery enables the physicist to *analyze* physical notions more precisely, and multiple and varied mathematical demonstrations enable the physicist to *understand* what is going on in the world ("the physics") since each demonstration reveals an aspect of the phenomena of concern.

The physicist's experience of mathematics is that mathematics would seem to do the bidding of physics: the formal objects the mathematicians create are wondrous tools for expressing and discovering "the physics," the mechanisms behind what we see. That mathematical machinery is customized and adapted for the work at hand, as is the case for all machinery. In part, some of the mathematics is developed because it is needed by the physicist; and physical concepts are often phrased so that they may be mathematically expressed. Of course, the world, the actual empirical world, or at least how we set it up in the laboratory, must in the end give substance to the mathematical machinery and the physical concepts.

On the other hand, it all depends on the available mathematics. If there is no mathematical technology, no tools suited to the job, the physics will not yield to mathematical investigation. (The physicist might hope to invent the technology, much as Newton, Gibbs, and Dirac did.) And, the suitability of the mathematical machinery, and its demand for rigor, is physically significant. Mathematical physics and its rigor is not just for show: that rigor and precision often reveals the essential physics. So, for example, a particular kind of mathematical convergence of a series of approximations to freezing or a phase transition points to that transition's distinctiveness.

"The physics" is a term of art, saying that what we want is to understand what is really going on in the physical process, and by "really going on" we are expressing the need for a deeper understanding, and by "deeper" we are seeking a mechanism or a more general theory. Similarly, "the mathematics" is used by the mathematicians about their own machinery. The machinery should not only work, but we should understand why it works, and from many angles or perspectives. The phenomenologists would say that we are searching for "an identity in a manifold presentation of profiles" – that apple – more general notions and a deeper understanding through many points of view. So, to solve a particular physical problem in many ways, or to prove a mathematical result in many different ways, is that manifold presentation of profiles, and that deeper understanding derived from those manifold presentations is the identity.

MACHINERY AND PRODUCTION PROCESSES

When machinery is employed in processes of production, it allows for greater efficiency, speed, and precision. What once demanded very great skill and took a long time now could be done with journeymen's skill and even be mass-produced. And the quality and kind of produced objects is intimately connected to the machinery that is available for producing those objects. Moreover, there are machinists who actually make those machines using other machines, and inventors and engineers who design all those machines and adapt machinery for new production processes. At any one time, there is a range of available machinery and modes of production, and what seems difficult and expensive to produce may well become easy and cheap in a generation. An archaeologist, whether of ancient societies or modern industrial ones, tries to describe the relationship of the machinery to the objects produced and the organization of society.

In this chapter we are in effect archaeologists, describing how mathematical technology works within a subculture. The machinists are the pure mathematicians, the production people are the physicists, and the inventors are either physicists or mathematicians who encounter a problem for which they might invent a method or machine, or perhaps they are mathematicians who like to tinker in their own workshops creating new and wondrous machinery (that is, doing mathematics) for which they have no particular non-mathematical use, although some designer might eventually employ it in making machine tools.

Machinery makes automatic what was once handcrafted work, although that machinery regularly needs to be tended by hand and fixed if it is to do its work reliably, not to speak of moments of breakdown. And so we, as archaeologists, might show why the materials and objects that people want are amenable to machinery, and why and how machinery is made adaptable for new materials and products. Moreover, we might inquire (as would an engineer) why different machines might produce the same objects, and why different kinds of objects are produced by the same machine. What makes the machinery or the objects generic? And how do people make the machinery work?

But now we return to how mathematical machinery works in the production of physics.

NAMING AND MODELING THE WORLD

If mathematics can have anything to do with the physical world, that world must be "named" or modeled mathematically – where by "mathematical modeling" I mean expressing a physical system in mathematical terms, the word "modeling" implying that the expression is schematic and incomplete. So, if we are to understand freezing using mathematical devices, we have to so express the notion formally and mathematically. The obvious fact that a liquid turns into a solid could mean an emergent orderly less-symmetric crystalline structure arising out of a liquid's disorderliness, or a change in density or compressibility, or a mathematical function having particular properties such as being zero at the transition-point temperature.

So, for example, it is the achievement of Willard Gibbs and thermodynamics to have found a way of expressing such transformations, namely as phase diagrams. A phase diagram is a geometric or topological statement about there being different regions of temperature-pressure space, matter being in a specific phase (solid, liquid, . . .) in each region, separated by well-defined boundaries. And the sharp transitions are here seen as discontinuities in density as a function of temperature, say, if the pressure is fixed. And there are mathematical rules about these diagrams' topologies (the Gibbs phase rule).

But, say you wanted to give a molecular constitution of matter, and working from the molecular level you wanted to express freezing. Freezing is a feature of bulk matter, not of a small collection of molecules. Only when you have a very large number of molecules is there sufficient interactions among them so that they march in lockstep. Then one might have a solid or a liquid. You need a way of transitioning from the molecular level to the bulk level (from 1 or 10 to 10^{23} molecules), and that transitioning must allow for finding liquids for higher temperatures while finding solids for lower temperatures, and at some intermediate temperature there is a discontinuity in the density, that of a liquid to that of a solid.

More generally, how do you build up bulk matter out of its constituent molecules so that, depending on the temperature or pressure, it is a gas or a liquid or a solid? This is a physical and a mathematical problem. Here we return to the discussion in chapter 2 about adding up the world. Namely, one might build up matter one molecule at a time, and with an infinitude of molecules one then has achieved bulk matter. Now, to make this statement practically useful for a proof or a derivation, you need a mathematical mode of building. One conceptual way of so building is to start out with a small cube of space, perhaps containing one molecule, and then making a larger cube out of eight of those cubes, and continuing the process.

Another way of so building is to imagine that space itself is divided into balls, and one then fits balls in the intersticies between balls and also fills up the balls themselves, in each case with smaller balls, say each ball roughly electrically neutral so it does not interact much with other balls, neither attracting nor repulsing them.[3]

Several problems remain: Can such a process be expressed mathematically? Does the mathematical building-up process end up with something that models bulk matter?[4] Can we deduce and prove properties of this notion of bulk matter from our molecular construction? Moreover, can such a construction accommodate the discontinuous transition from liquid to solid at the freezing point – especially given that as we build up bulk matter we are doing so incrementally?

How could discontinuity arise from an incremental process? What the mathematicians offer the physicist is the fact that some building-up or limiting processes, using incrementally changing behavior, such as increasing the number of molecules, models used to build up bulk matter, need not lead to a smoothly changing function in the limit.[5]

Another naming or modeling problem concerns apparently smooth continuous processes of diffusion and flow. Consider describing mathematically what happens when we open a perfume bottle. After a short while the perfume's fragrance would seem to spread throughout the room. Presumably, perfume molecules evaporate from the liquid, which molecules then seem to find their way to every corner of the room. Such diffusion processes may be modeled mathematically and physically by the random walk we described in chapter 2, where at each moment a walker chooses arbitrarily to go one step north or south or east or west: over time the walker is most likely to be found at a distance from its origin point, that distance proportional to the square root of the number of steps. Namely, at each moment a perfume molecule may bump into another molecule, most likely a molecule of ordinary air, and be sent off in a different direction. In fact, this picture or model is a good one for describing what really happens. We say that the fragrance molecules *diffuse* with an average distance proportional to the square root of the number of steps or collisions they make in that time. Hence you might well time how long it takes for a perfumed lady, upon entering a room, to fill that room with her fragrance, in order to find out about molecular collisions. We say that these molecular processes involve molecular collisions characterized by a rate of collision and so a mean distance between collisions, whose collective effect is diffusion. There is a nice connection between the individual collisions and the rate of diffusion, and so we might study microscopic processes of molecular collisions by studying the macroscopic process of the diffusion of the perfume and its fragrance – and the precise connection between the two is the temperature (higher temperatures mean more rapid diffusion).

Now, such fluidic or gaseous motion need not be characterized molecularly. For example, we might say that the change of the amount of fluid in a small volume of space is the difference between the inflows and the outflows over a fixed period of time. Vectors are the natural expression of the flows since they encompass both direction and quantity, while ordinary numbers are useful for expressing the quantity of fluid in that small volume of space. And the calculus is well known to be a good way of expressing changes. What is needed is a machinery that combines the calculus with vectors, one that would be useful for expressing flows of a material. The vector calculus (in the symbolic form due to J. W. Gibbs and E. B. Wilson's *Vector Analysis,* 1902) is a language that expresses how a quantity changes in space or time, how a flow accumulates or dissipates in a place, and how rotational motion or circulation of a fluid affects other of its properties. The mathematical formulation also shows the fact that for a steady state, not changing in time, the number of molecules in a small region of space is an average of the surrounding small regions' numbers.

What is notable is that that same vector calculus is also able to express electromagnetism, where the fluid is for the most part not a flow of charge but a flow of "fields" such as the magnetic field. (Recall chapter 1 on fields and particles.) Treating fields as flowing fluids, notional fluids to be sure, the mathematical machinery for fluid flow proved amenable to electricity and magnetism and their interactions. It is also the model for field theories of elementary particles.

An archetypal question then arises, why do two very different mathematical machineries or dialects work in a single situation (here, that perfume's diffusion): random walks and the vector calculus? How and when are a random walk and a flow like each other; how can the collisions of the perfume molecules be seen as a flow? (The answer in this case is that we might average the number of molecules in each small volume and so go from individuals to a flow.)

In each case – freezing, bulk matter composed molecularly, diffusion, and fluids – mathematical machinery (limit processes, random walks, vector calculus) is employed to say more precisely what we mean: by a phase transition, by bulk matter, by diffusive flow, and by a fluid or a field considered as a fluid. Those devices enable further exploration of the properties of the physical world through mathematical manipulations – although some particular properties are surely left out of or at least not made manifest by one set of mathematical machinery and are better approached through different mathematical machinery. So, for example, efficient devices for counting up molecules may not make it easy to discern orderlinesses that emerge in the collection of molecules.[6]

Employed carefully, the mathematical machinery would seem to by the way do lots of the work of physics, for free. That is what we might hope from machinery. Moreover, that scheme and automaticity shapes or names how we think about the world. Physics shows new uses for old machinery, finds new machinery, and shows how Nature is a product of, is nicely modeled by, convenient mathematical machinery that turns out to be physical as well. This is no miracle, but reflects an enormous effort to conceive of the physical world in ways that are mathematical, to invent mathematics that is needed to model Nature, and the efforts of production people to make that machinery work.

DEMONSTRATIONS AND PROOFS AS STRATEGIES OF EXPLANATION

In our ordinary everyday experience, matter is stable. That "[a] stone is solid and has a volume which is proportional to its mass, and . . . bringing two stones together produces nothing more exciting than a bigger stone," is in fact a consequence of quantum mechanics. And molecularly, the fact that matter does not collapse, in effect the electrons and nuclei falling into each other, and it does not fly apart, the electrons repulsing each other and so flying apart from each other, and the same for the nuclei, is also such a consequence. In 1939 Lars Onsager provided an initial albeit rough physicist's proof of the stability of ordinary matter. But Onsager did make some sensible and true and convenient but not proven assumptions, ones that demand quantum mechanical proof.

Technically, one has to prove that a one-electron hydrogen atom is stable, and that a multielectron atom (such as carbon) is stable, each having a small but finite size. Then you have to consider matter composed of many atoms, and show that it is stable, and that it is bulky as well. What's crucial is an uncertainty principle, that electrons obey the Pauli Exclusion Principle, and that the positive charges of the nuclei are "screened" by the atomic electrons, so that their effect is not much felt a few atoms away. In effect you are demonstrating that "bulk matter is stable and has a volume proportional to the number of particles . . . [because] the electrons behave like a fluid . . . and this limits the compression caused by the attractive electrostatic forces."[7]

One of the notable achievements in mathematical-physical explanation was the rigorous 1967 proof by Freeman Dyson and Andrew Lenard that ordinary matter is stable. The challenge here is to prove that stability is the case given what we know about atoms and molecules, that they are composed of charge-bearing electrons and nuclei, and about physical laws (electromagnetism, quantum mechanics).

There are two challenges here. The first is to find a good formal mathematical definition of what we ordinarily mean by "stable," that is, a mathematical abstraction that we believe encapsulates what we mean by "the stability of matter," the naming problem. And the second is to use the physical laws, as they are expressed mathematically, to prove that ordinary matter is in fact stable. A good mathematical proof must be valid, with no lacunae or errors, and it also must be convincing and illuminating, along the way telling us why it works and so showing us why it is valid. Knowing the laws and having a mathematical model of stability is not enough: how are they to be deployed to make a proof? The Dyson and Lenard proof was at the time one of the longest, most complicated, and ingenious arguments in physics and in mathematical physics. The authors do provide insight, showing for example that it is crucial that electrons are particles each of whose labels or quantum numbers are unique (called "fermions" and so obeying the Pauli Exclusion Principle), so you cannot have two in the same place. If electrons were not uniquely specified ("bosons") the proof would not only not go through, but one could prove there could not be stability. And they arrange the proof in such a way that one can see how it is structured in physically sensible terms as well. In effect one has lemmas hanging from a tree of theorems, that tree having some "warm-up" arguments and theorems, then a toy problem of just one electron particle in a sea of charge, and then the proof of stability of matter using these lessons but now for many electrons. The machinery works, and we can better understand how it works, and so why matter is stable. Still, one might understand each step, but not find the proof really illuminating. (Dyson referred to it as hacking through a forest of inequalities.)

Subsequently, there were mathematical proofs that were simpler, and, eventually, where the relevant physics was highlighted and more readily seen. The original proof was supplemented by a cleaned-up, more-mathematical proof by one of the original authors, then by a proof that used the mathematical and physical technology of quantum field theory, then by a quite physical mathematical proof that capitalized on a model of the atom that averaged the electron cloud around an atom (the "Thomas-Fermi model"), and then by formally more rigorous mathematical proofs, and then by a sequence of mathematical improvements employing sharpened mathematical inequalities that, too, sometimes had physical import. It is surely crucial that electrons be fermions, and that a single atom itself be stable (a quantum mechanical uncertainty principle). But, it is probably the case that the Thomas-Fermi proof was most revealing of what was going on: not only was the positive nuclear charge hidden or shielded by the negative electrons surrounding the atom, and correspondingly the electrons' effects were neutralized by the nuclei, but in effect that hiding or screening was for our purposes absolute.[8]

As another example of how the physics emerges from varied derivations or proofs, each revealing an aspect of the object, consider the first mathematical demonstration of a phase transition in matter (an iron bar becoming permanently magnetizable) starting out with molecules and their interactions, by Lars Onsager in 1944.[9] (See the preface and Figure P.1.) It is a demonstration notorious for its length and complexity and its use of then arcane mathematical objects. One could discern a structure of proof, first setting things up so that what one is doing is solving a problem much like the first-quantum-mechanics-course problem of the quantum harmonic oscillator – where the kinetic energy and the potential energy terms are both of comparable importance.

In the many subsequent proofs over the next sixty years, the crucial physics became clearer, and the mathematical technology employed was often of a more familiar sort and so it was more readily followed. Eventually, one could figure out just why the phase transition could take place, and how Onsager's proof built in some aspects of the crucial physics, even if that was not so apparent on an initial reading: the "particles" in the system (those right degrees of freedom) were regular up-and-down arrangements within a row of molecules, where each arrangement considered as a particle behaved like a fermion (and an infinite number of them, each having almost zero energy, were "excited" at the phase transition point, and so the specific heat, that might indicate a phase transition, was infinite at just this point), and in fact there were pairs of fermions as in superconductivity; and, there were various scaling symmetries of the orderliness of the lattice that were characterized by good variables, good degrees of freedom, that *just happened to be* like the variables in the trigonometric or circular functions (sine θ, cosine θ, . . .), but here they are "elliptic functions," having both shape and angle variables, where the variable that characterizes how elliptical is the shape also characterizes the scaling symmetry, as well as characterizing the relationship of a high-temperature disorderly state that is not magnetizable to a low-temperature orderly magnet. Of course, we would want an explanation of how that "just happened to be."

The mathematical machinery surely has to work. But it also must be set up so that we are convinced that there are no significant lacunae, *and* that we understand why and how the machinery does its work. Along the way, physical insight is often helpful. What is crucial is not simply the initial proof or demonstration (*tour de force* that it is), but that it be understandable, and the understanding provided by such a mathematical display leads to a physical understanding of why the world is the way it is – so that the mathematics may be put aside and one's physical intuition allows one to understand what is going on. The machinery is instrumental, allowing us to better understand the workings of the world.

The mathematics is a form of philosophical analysis: rigorous, detailed, clear, and it provides an argument (the proof). Moreover, in clarifying the argument, and in dealing with the niceties of mathematical rigor, physical features of the system are highlighted or revealed – so that details of the orderliness of that Ising matter become manifest. Moreover, the multiple proofs provide understanding.

UNDERSTANDING "THE PHYSICS"

In the end, what you want is to understand what is really going on, why something is the case. And, as I suggested earlier, different mathematical demonstrations will show different features of the system under study. Usually, the earliest demonstrations are awkward, complex, lengthy, and perhaps obscure. Again, each is a *tour de force,* as was the Onsager paper of 1944 or the Dyson-Lenard paper of 1967. Subsequent demonstrations may simplify one part of the argument. But what is most revealing is a demonstration that employs novel concepts, strategies, or mathematical machinery, and then follow-on demonstrations, so that we see that the original quite complex demonstration is not merely machinery that just happens to work, but that it manifestly and actually builds in some of the physics (the mechanism, the physical processes) of the system. In retrospect, we may even appreciate its elegance. Moreover, each demonstration provides a different perspective on the system, so that we have several ways of thinking about the system and so understanding why it is the way it is.

Again, Onsager's original 1944 proof can now be seen to have many of the physical features revealed in the subsequent often "simpler" proofs, those perhaps using more familiar mathematical machinery. Even more revealing, the set of proofs illuminate each other, indicating why Ising matter behaves the way it does and why each of the various diverse mathematical machineries might actually work. The machineries point to different physical features, which we might then see as a complex of related features of this sort of matter – Ising matter accommodates various different aspects, much as does an apple. Here might be an identity in a manifold presentation of profiles.

We might classify these many proofs, and there are many of them, into three categories: arithmetic, algebraic, and analytic (the latter as in the calculus, nothing to do with philosophic analysis). The arithmetic proofs depend on literally counting up the interactions of the adjacent molecules, perhaps employing matrices to do that counting – matrices that when multiplied "automatically" count up the interactions. Those counting matrices actually reveal some symmetries of Ising matter – for example, the relationship of properties of high-temperature Ising matter to those of a corresponding low-temperature state ("duality" is the physicists' term of art). Algebraically, those matrices are not merely devices. For example, one can show that the matrices reflect the fermion-nature of the "particles" (those orderly rows) in Ising matter. In any case, once one has the matrices, it turns out that their algebra, and then analytic features of the solution, are needed to solve the problem. In other words, there is a good deal of overlap among these three categories.

Algebraically, one can show that one kind of such counting matrices (not quite the ones in the previous paragraph) may interchange or "commute" with others of the same kind, that is $TrHx = HxTr$ – where Tr and Hx here represent counting matrices for a triangular and a hexagonal lattice that both have the same degree of orderliness, albeit they are systems at different but low (or high) temperatures. (Onsager labeled this the "star-triangle relationship," referring to the hexagonal and the triangular lattices, respectively.) The good degree of freedom here is not the temperature but a measure of the orderliness. What appears to be a geometric or topological statement actually points to those variables in the elliptic functions, in effect analytic features, again one of which is a measure of orderliness or the shape of an ellipse. The algebra, that the matrices commute with each other, is physically meaningful, and it turns out to allow one to solve the Ising-matter problem in a very different fashion than employing the arithmetic machinery.

Analytically, one makes a direct assault on a solution by incorporating all the orderliness that has been discovered into a proposed solution – where one makes some reasonable assumptions about the smoothness of the solution, and assumptions that depend on knowing propeties of a solution, such as where the zeroes are (so for the sine function, they are at 0, π, 2π, . . .). And then, by what might be called an ingenious legerdemain, one solves the problem using what we know about functions of a complex variable (treating apparently real numbers as complex numbers), a highly constrained environment.

As I have indicated, there are different systems of matrices, at least from the arithmetic and the algebraic solutions, whose properties (their "traces") provide solutions for the properties of Ising matter. And the arithmetic, algebraic, and analytic features of the solutions can be attached to parallel mathematical features of what is in effect a doughnut or torus (what is called an "elliptic curve"). I mention these extra parallels since they will be important shortly.[10]

Again, each of the mathematical methods and machineries, except perhaps the first arithmetic solution by Onsager (and he would seem to have learned from earlier partial solutions by others), has depended on knowing what was revealed by earlier methods. That does not discount the value of alternative approaches, for each vividly displays different important features of Ising matter.

Technically, we might say that Ising matter allows these various descriptions since it is the product of local interactions of the atomic or molecular magnets, and differently configured lattices end up behaving the same macroscopically or in bulk, and there is a pervasive scaling symmetry from the microscopic to the macroscopic. Ising matter accommodates all these aspects, at the same time. Ising matter is that identity in a manifold presentation of profiles. And, each kind of proof – arithmetic, algebraic, and analytic – might be employed to show how the other types may be derived from it.

But, what we might say is still insufficiently satisfying. Why do these three types or characterizations all work? Where is the physics that grounds this identity? How are we to think of Ising matter so that its features are deeply connected, and not merely a list connected by methods of mathematical demonstration? That locality, those relationships, and those symmetries should be intimately connected. How do these three properties go together, and why the matrices and their properties? Might what would appear to be arbitrary devices and tricks employed for solving a problem (they worked!) be more deeply connected and point to a deeper source?

ANALOGY AND SYZYGY

That Ising matter might be understood arithmetically, algebraically, and analytically and scale symmetrically is an analogy. But, again, why are there these three moments of the analogy? Why the matrices? Now, there is another partially proven but otherwise merely conjectured analogy, but now in pure mathematics, associated with enumerating the prime numbers. Again, there is an arithmetic or combinatorial method, there is algebraic method, and there is an analytic (and geometric and topological) approach. And, again, there are several systems of matrices (what are called "group representations") which embody these features, and their properties (traces, again) provide solutions for the various approaches. And the doughnut also appears, as a model of these relationships, again its having the appropriate arithmetic, algebraic, and analytic properties. What is hard to demonstrate about the prime numbers in one approach or moment of the analogy is easy in another: hence, algebra can help arithmetic, and analysis can help algebra. This analogy has been called a Rosetta stone, allowing for a translation of notions from one moment of the analogy to another.[11]

What is striking is that one might set up a tentative analogy between the physicists' analogy and the mathematicians' analogy, what is called a syzygy (an analogy of analogies). Here, what the mathematics provides is a potential reason why the physicists' analogy has the moments that it does have. And the physics provides a concrete model of the conjectured mathematical analogy. Now, the big question remains: Why are there these three aspects to these systems? In other words, What are the deeper foundations for either of the analogies or of the syzygy?

THE MATHEMATICS AND THE PHYSICS

Mathematical machinery may provide the language of physics, as Galileo suggests. That language works in various ways – allowing for more precise expression of our notions, enabling us to demonstrate the properties of physical systems and why they have those properties, and it may lead to an even deeper understanding through the various ways we might understand a system. The mathematical machinery would seem to have an internal logic which then is expressed in its account of the physical world. Presumably that machinery was developed in part because it proved useful, and the physics developed in part because there was language and machinery to express the actual world.

Apropos of mathematical machinery to express the actual world, there is a tradition of physicists hunting for an equation or some mathematical machinery having the right features to do the work. So, Schroedinger found an equation when Debye challenged him to find one that would fit deBroglie's informal insights – for their Sommerfeld culture demanded that there should be a differential equation if it was to be good physics. Heisenberg developed an apparently artificial formalism that incorporated what was known about atomic spectra, and very soon after others realized that it was a matter of matrix multiplication. Dirac found his equation given some reasonable desiderata from special relativity. Einstein's path to the equation of general relativity was a matter of physical and mathematical demands, which then became requirements once the equation was found.

Along the way, physicists may find equations and machinery that do not work. So they continue the hunt.

SEVEN

An Epitome

PHYSICISTS' THEORIES AND PICTURES OF NATURE ARE ANALOGIES with everyday phenomena and objects (hereafter, "objects") such as a factory, a mechanism, or family relationships. How physicists get hold of the world through those analogies *is* the way Nature is for them. They understand each analogy in a highly stylized and particular way, and to so understand Nature is to be trained as a physicist.[1] When understood in one of those highly stylized ways, the physics, the everyday objects, and the grammar or language overlap nicely. Nature would seem to accommodate many perspectives or many such analogies, and a particular object or phenomenon may be adequately described by more than one.

What physicists want is a deeper understanding of what is going on. The analogies allow them to name features of Nature and analyze them. Getting hold of an object in multiple ways, leading perhaps to analogies of analogies, can lead to that deeper understanding. In order to name and understand an object by analogy, physicists isolate features of both the physical object and the analogized everyday object, each at an appropriate scale or size of energy or length, while they ignore most other features. In other words, parts represent the whole: in rhetoric called a synecdoche, in anthropology called fetishism. Such fetishism, modeling Nature, has proved to be an effective strategy.

Our everyday understanding of objects influences how physicists understand objects in Nature. And physicists' understanding of their objects affect our everyday notions – again, a theme in the history of ideas and of social science.

Good degrees of freedom are handles onto an object that allow us to probe that object, subtly and marginally, and sometimes not so subtly and qualitatively transformational, by pulling on or shaking those handles. Good degrees of freedom are in effect strategies for getting at the world. Some degrees of freedom are mostly hidden, handles we cannot get hold of. Yet we see their effect through fluctuations. Others are effectively totally hidden.

We have discussed six sources of everyday analogies: the factory and its division of labor; a mechanism composed of parts in interaction; the family with its marriage or exchange rules, where everything that can happen must happen; the theater, which provides a simplified stage on which there is both order and symmetry, Something and Nothing, an account of creation; the physicist's toolkit of handles, probes, and tools, where objectivity and observation are physical processes; and the machine shop, where mathematical machinery and skilled workers (physicists and mathematicians) attending to that machinery allow that machinery to work in seemingly automatic ways to produce physicists' account of Nature. Our recurrent substances have been ordinary everyday bulk matter (gasses, liquids, solids), microscopic elementary particles and fields, and Ising matter as a model of a magnet made up of microscopic magnetic molecules in interaction. Our recurrent theories have been thermodynamics and field theories (such as electromagnetism and quantum field theory).

Our goal in *Doing Physics* has been to provide a description of how physicists get hold of Nature, a description that might seem right and obvious if not usual to a physicist, but which is also familiar to laypersons, part of their ordinary everyday experience.

Notes

Some of these notes are for the lay reader; others are for the expert. I have also provided elaborations that will be useful to the student, but they may well seem superfluous to the professional. As indicated in the preface, the notes often provide technical justification and details for what I say rather more ordinarily in the text. In this manner, I would hope both to indicate that I am aware of some of the usual objections and to say more precisely what I mean, in order to avoid misreading by those concerned with matters technical.

As will become apparent, I have employed a number of sources almost as texts, namely: an essay on degrees of freedom and the renormalization group by Steven Weinberg; a book on condensed matter physics by P. W. Anderson; J. A. Wheeler on fields and sources; Landau and Lifshitz on statistical physics; and Feynman on more than a few things. My book *Marginalism and Discontinuity: Tools for the Crafts of Knowledge and Decision* provides background for the more methodological claims made herein.

A number of examples, models, and notions have appeared recurrently in the text. In these topics, and in their recurrence, lies the message of this book.

PREFACE

1. C. P. Snow, *The Two Cultures and the Scientific Revolution* (New York: Cambridge University Press, 1959). I am not claiming that there is just one universal culture. Science is differently practiced, in important ways, in different cultures. For a comparison of the United States and Japan, see S. Traweek, *Beamtimes and Lifetimes* (Cambridge, Mass.: Harvard University Press, 1988). The interesting question is to account for the sharedness of the work of science in different cultures. But I think this is no more and no less problematic than the tasks of comparative religion or comparative literature. See also note 10 below.

I should note that one of the enduring themes in the history of science is the connection between economic-political systems and the structure of scientific theory. The locus classicus is perhaps B. Hessen, "The Social and Economic Roots of Newton's 'Principia,'" in N. Bukharin, *Science at the Cross Roads* (1931; reprint, London: Cass, 1971). Obviously, important work in physics has been done under socialist or communist influence. See, for the example of V. Fock in the Soviet Union, L. Graham, *Between Science and Values* (New York: Columbia University Press, 1981). And much of contemporary talk about "chaos" seems to indicate the importance of collective modes. Yet, what is curious is how those modes often become individual particle-like objects.

2. J. C. Maxwell, "Thompson and Tait's Natural Philosophy," in vol. 2 of *The Scientific Papers of James Clerk Maxwell* (Cambridge: Cambridge University Press, 1890), pp. 782–84, partly quoted in J. Z. Buchwald, *From Maxwell to Microphysics* (Chicago: University of Chicago Press, 1985), p. 21. I follow the 1890 text.

3. N. Frye, *Anatomy of Criticism* (Princeton, N.J.: Princeton University Press, 1957). See the seminal G. Holton, *Thematic Origins of Scientific Thought* (Cambridge, Mass.: Harvard University Press, 1988) for a historical analysis. The quote on conventions and craft comes from S. Shapin and S. Schaffer, *Leviathan and the Airpump* (Princeton, N.J.: Princeton University Press, 1985), p. 18, where they refer to S. Alpers, *The Art of Describing* (Chicago: University of Chicago Press, 1983).

4. S. Weinberg, "Why the Renormalization Group is a Good Thing," in A. Guth, K. Huang, and R. Jaffe, eds., *Asymptotic Realms of Physics* (Cambridge, Mass.: MIT Press, 1983), p. 16.

5. J. Clarke, A. N. Cleland, M. H. Devoret, D. Esteve, and J. M. Martinis, "Quantum Mechanics of a Macroscopic Variable: The Phase Difference of a Josephson Junction," *Science* 239 (26 Feb 1988), at p. 997.

6. For an interesting analogy, see H. Adams, "The Rule of Phase Applied to History" (1908), in E. Stevenson, ed., *A Henry Adams Reader* (Garden City, N.Y.: Anchor, 1958).

7. If we have described a system fully in terms of its constants of motion (or conserved quantities), then that "change" is in the configuration of the system (not a change in time, as such) – say, a system at a different energy or angular momentum.

8. It has been suggested to me by Sherry Turkle that what I am up to is what James Strachey, the psychoanalyst, has called a mutative interpretation, a description of what we are doing that integrates what seem like disparate parts and so leads to a transformative self-recognition.

9. You need two dimensions to get enough connections among the atomic magnets to get bulk magnetization, that is a magnetized piece of iron – one dimension as in a line of atoms is not enough connectivity. Also, up and down is an abstraction from any possible direction.

10. S. S. Schweber has pointed out how American and perhaps post–WW II is the culture I describe, especially in the willingness to take the world as it is analogized and not worry about deeper meaning, at least provisionally. "Favoring practice over theory, action over thought, invention over contemplation has been called an American trait." Analogy and tools are fundamental to this pragmatic point of view. The current practice of "effective field theories" is in this vein (but cf. string theory, which also has strong American origins). See S. S. Schweber, "The Empiricist Temper Regnant: Theoretical Physics in the U.S. 1920–1950," *Historical Studies in the Physical Sciences* 17 (1986): 55–98, at p. 59. There are other more speculative cultures in physics, as in parts of Europe during the twentieth century, which have been productive, to say the least, yet not quite so willing to take the world just as it is analogized.

My concern here has not been to argue for operationalism and the like. Rather, I want to describe the modalities by which physicists actually do physics.

11. T. S. Kuhn, *The Structure of Scientific Revolutions* (Chicago: University of Chicago Press, 1970), and *Black-Body Theory and the Quantum Discontinuity* (New York: Oxford, 1978); I. Hacking, *Representing and Intervening* (New York: Cambridge University Press, 1983);

D. Gooding and T. Pinch, eds., *The Uses of Experiment* (New York: Cambridge University Press, 1988).

12. The analogies that concern me here are neither necessarily formal nor material, to use Hesse's terminology, although they sometimes are. M. Hesse, "Model and Analogy in Science," in vol. 5 of the *Encyclopedia of Philosophy* (New York: Macmillan, 1967), pp. 354–55.

13. A. Stephanides and A. Pickering, "Constructing Quaternions: On the Analysis of Conceptual Practice," in A. Pickering, *Science as Practice and Culture* (Chicago: University of Chicago Press, 1992). This article works out the process of extension in some detail; the authors draw from Kuhn and Hesse among others. My interest here is to catalog the major models.

14. See P. Brown, chap. 23 of *Augustine of Hippo* (Berkeley: University of California Press, 1969).

15. For example, in the text I pay comparatively little attention to some of the apparent peculiarities of quantum mechanics. This was deliberate. In the subsequent years, quantum coherence (and Bell's inequality) has become an industry. Yet, in the notes I believe I have dealt with these issues adequately. I remain committed to physics as being practical and commonsensical (if you have well-trained common sense!), not at all mysterious.

16. D. Pines, "Superconductivity: From Electron Interaction to Nuclear Superfluidity," in L. N. Cooper and D. Feldman, *BCS: 50 Years* (Hackensack, N.J.: World Scientific, 2011), pp. 85–105, at p. 103. Also, S. S. Schweber, *Nuclear Forces: The Making of the Physicist Hans Bethe* (Cambridge, Mass.: Harvard University Press, 2012) provides a similar strategy, albeit of another "off-scale" scientist.

1. THE DIVISION OF LABOR

1. Of course there are many different political economies of the transcendental aesthetic. See, for example, J. Z. Buchwald, *From Maxwell to Microphysics* (Chicago: University of Chicago Press, 1985), for an account of the Maxwellian interpretation of field theory, one in which fields are "all" there is, and a comparison with the modern interpretation in terms of fields and sources. See also J. A. Wheeler, chap. 6 in *A Journey into Gravity and Spacetime* (New York: Scientific American Library, 1990), on the "grip of spacetime on mass" and vice versa.

2. Adam Smith, *The Wealth of Nations* (1776) (Chicago: University of Chicago Press, 1976).

3. For a review of this tradition, see L. Dumont, *From Mandeville to Marx* (Chicago: University of Chicago Press, 1983). On stickiness, see M. H. Krieger, chap. 4 on "Sticky Systems," in *Marginalism and Discontinuity: Tools for the Crafts of Knowledge and Decision* (New York: Russell Sage Foundation, 1989).

4. D. Hume, in pt. 5 of *Dialogues Concerning Natural Religion* (New York: Penguin, 1990), p. 77.

5. In biology, this is often quite explicit. See, for example, M. Ghiselin, *The Economy of Nature and the Evolution of Sex* (Berkeley: University of California Press, 1974).

6. Correspondingly, we might ask how do we delineate time so that events follow each other in a regular way (as in linear differential equations)? The dynamical principles, which we discuss at the end of this chapter and in subsequent chapters, are such delineations.

7. Shortly, I shall point out in the text that I am not concerned here with the division of labor (as in a laboratory) in the production of physics. What we are concerned with is the production of Nature, a distinction which might well subject me to a criticism that I hypostasize Nature. I do take Nature as that something which the physicists understand quite unprob-

lematically, even if it can be shown that Nature is a product of their theorizing and social construction.

8. There is, as well, an essential ambiguity, about which I say little here: what Nature is, the design we impute, and whether there is a designer. But see M. H. Krieger, "Temptations of Design: A Meditation on Practice," in *Research in Philosophy and Technology* 10 (JAI Press, ed. F. Ferré, 1990): 217–30, reprinted as part of "The Manufacture of the Sacred, the Reenactment of Transcendence, and the Temptations of Design," in M. H. Krieger, *What's Wrong With Plastic Trees? Artifice and Authenticity in Design* (Westport, Conn.: Praeger, 2000), pp. 75–94.

Of course there might well be analogies between the material world of scientific organization and the ideational world of scientific theory. The analogies that I provide are between the material world of everyday life and the ideational world of scientific theorizing.

9. Imagine individuals that were delineated by bell-shaped or Gaussian distributions so they overlapped. Even though there is no obvious wall that separates them, our conventions about "cutoffs," and about fitting the sum distribution to two Gaussians, allow us to enclose each by a wall (for example, by its mean and standard deviation, the means and/or standard deviations perhaps being different).

10. Harrison White has described interfaces in social science contexts in terms of the interdigitation of systems and the matching of their variances. See H. White in *Connections* (Toronto) 5 (1982): 11–20.

11. As a consequence of that independence, energies are additive, for example. See L. Landau and E. Lifshitz, secs. 1, 2, and 4 of *Statistical Physics* (Oxford: Pergamon, 1980).

12. The chemical valence or bond is also such a skin.

13. P. W. Anderson, chap. 2 in *Basic Notions of Condensed Matter Physics* (Menlo Park, Calif.: Benjamin/Cummings, 1984), especially pp. 49–59, provides a lovely classification of what I have called walls: Walls are marked directions in otherwise symmetric arrangements, breaking up that symmetry. I should note immediately that Anderson is concerned with walls considered as defect structures (pp. 52–53), such as domain walls. My use is rather in the sense of a shield, but a shield is just what permits the conveying of a symmetry across a boundary (actually a hiding, of course).

According to Anderson, there are three kinds of broken symmetry. A crystal lattice breaks up the homogeneity or symmetry of space; and it allows for collective excitations or vibrations of the whole lattice, that is, the emergence of new particles such as phonons. (Elementary excitations, such as quasiparticles, represent a very different principle, that of "adiabatic continuity," electrons smoothly going into quasi-electrons in the lattice.) Another kind of broken symmetry is associated with broken gauge symmetry, and hence the W and the Z particles. And a third form of broken symmetry is broken time reversal invariance, from which we get the possibility of permanent magnetization. In each case, one finds the following: discrete phase transitions (sharp separations, just what walls are expected to do); collective excitations or emergent properties (long wavelength excitations of the order parameter, namely fluctuation and permeability of the wall as, for example, lattice vibration); "generalized rigidity" (or long-range order, as we might expect for a wall); and defects in that rigidity and hence consequent dissipation.

Not all phase transitions are broken symmetries as described above. Phase transitions can also be a matter of nonlinear instability; or, as in percolation, there is a conformal invariance at the transition point.

Technically, we might say that particles are shielded from outside by a surface that is a consequence of a generalized rigidity due to a phase transition, a surface that is of finite rather than infinite thickness, and which fluctuates but is stable. A surface is a shield in that it is impermeable to a class of insults, the enclosed particles being untouched by these insults, the particles being invariant to them.

14. An example: Technically, potentials hide and filter out many degrees of freedom, revealing or producing just a few (the good handles) – whether the potentials are thermodynamic (energy, free energy, . . .) or electrical ($-Q/r$) or gravitational, for example. Here, the hidden degrees of freedom are in fact *most* of the degrees of freedom of the gas particles, or of the electromagnetic field (more than one photon exchange). If there are nonuniformities inside a block of matter or inside a charged particle, what we see first is the mass or charge (representing uniformities), and then usually dipoles and quadrupoles which represent the nonuniformities in a still highly summary fashion.

Another kind of wall is the "thermodynamic limit," the existence of bulk properties such as temperature and pressure, and thermodynamic stability, so that matter becomes denser when compressed and the temperature goes up when you heat an object.

15. Technically, there is an exchange of momentum between the gas and the balloon. Similarly, to maintain a temperature there is an exchange of energy with a wall. And, in general, arrangements are drowned out because the walls are rigid enough that they do not respond strongly quickly enough to such transients (in effect they average and damp) so that we might see the transients externally. Sensors or probes cannot be too sensitive if they are to display average properties.

16. The orderliness provided by the regular arrangement breaks up another symmetry, a homogeneous isotropic arrangement of the molecules as in a gas, and so there is a now a wall. (See note 13.) And that orderliness also implies a residual ignorance (or symmetry) of the individual atoms that occupy the regular places. Namely, peculiar or idiosyncratic details at scales much smaller or larger than the crystal's lattice constant (or repetition distance) are comparatively hidden from us, and hence a no-matter-how. But, see the next note. And if we do X-ray crystallography, *regular* small details (such as those of the atoms that make up the crystal, or defects) are magnified because of that regularity.

17. Of course, if we shine in very energetic particles of light, we can see individual atoms. We have in effect broken through the wall.

18. Put differently, by definition of "elementary," an elementary object should not interact with itself. More technically, this becomes a problem concerning the self-energy of a particle, the energy of interaction of itself with itself (which we might expect to be zero since by its integrity it cannot interact with itself). Of course, the object or particle might be complex and composite when viewed at another scale. But at the scale for which the particle is taken as an individual, it should not interact with itself. See also the next paragraph in the main text.

19. Namely, adiabatic processes and other reversible transformations employed to enable the comparisons.

20. In electrodynamics, we say the walls conserve charge, allow for an infinite range force (namely, an inverse square force) so a particle can sense everywhere – and hence, as we discuss, the mass of the photon is zero, and hence

gauge invariance, and hence charge conservation – and be finite in thickness, not going to infinity ("renormalizability"). See L. Landau and E. Lifshitz, sec. 16 of *Classical Theory of Fields* (Reading, Mass.: Addison Wesley, 1962), on the connection between gauge invariance and charge conservation in the classical formulation. Again, the connections between the technical and the phenomenological are quite intimate. For, even more technically, in electrodynamics own-or-other means being able to sense everywhere (mass of photon is zero); which means there can be good charges (conservation of charge) with a workable division of labor between fields and particles, which means that the object is path-independently defined (gauge invariance); which means there are procedures for adding up its fluctuations (renormalization) so that in the end they only count in a very general consummate perturbative way. I use the term "means" in the previous sentence to suggest both a phenomenological analysis connecting these features (as in the text) and the usual mathematical-physical argument ("deduction" is too formal a notion here) found in textbooks. See note 52. Note that, by device, apparent nonzero masses and finite ranges are allowed in modern nonabelian gauge field theory.

21. In thermodynamics, for example, the specific heat is strictly greater than zero if the temperature is greater than absolute zero. For further details see chapter 4. See also note 22.

A deep question is why there is so little energy fluctuation in the physical vacuum, the problem of the small size of the cosmological constant. Why is there nothing rather than something? See S. Weinberg, "The Cosmological Constant Problem," *Reviews of Modern Physics* 61 (1989): 1–23.

22. This is a story of the fluctuation-dissipation theorem and, again, of renormalization. If a wall is supposed to keep a system at constant volume and temperature, the system will still have fluctuations in its energy, and those fluctuations determine how difficult it is to heat up the system (that is its absorptivity or specific heat). ($\langle \Delta E^2 \rangle \approx kT^2 C_v$; $C_v \approx Nk$; therefore, $\langle \Delta E^2 \rangle \approx N(kT)^2$, where kT is the energy per degree of freedom – as we might expect in adding random variables: where ΔE measures the fluctuation in energy, k is Boltzmann's constant, T is the absolute temperature, C_v is the specific heat, and N is the number of degrees of freedom. See E. Schroedinger, *Statistical Thermodynamics* (Cambridge: Cambridge University Press, 1964), pp. 25, 40. See chapter 5, note 35 for further discussions of fluctuations and permeability.

23. For example, in melting, the two separated phases become one, becoming "infinitely" intermingled. In demagnetization, we get an intermingling rather than a clumping in the pattern of the spins. And if an object such as a particle is thought to be a condensate of properties – namely, that a set of more global properties is in effect condensed out onto it – then when there is melting, that condensate no longer holds together and hidden degrees of freedom appear. See chapter 4.

24. See note 13. Technically and speculatively, sometimes we might even say that the physical mechanisms of internal constraint are literally the shields to the outside, those physical mechanisms preventing our seeing the held-in degrees of freedom: (1) The self-reinforcing interaction of electrostatic forces which keeps the molecules of a crystal in their symmetric places then limits what we can see electromagnetically (within a range of wavelengths) of this crystal (a mechanism that expresses redundancy). (2) The conservation of energy and of the various charges (which in effect tempers the de-

grees of freedom expressed in a quantum field) determines which fluctuations can show themselves. Quantum fluctuations of energy (which only appear to violate the conservation of energy) might have a size and duration given by $|\Delta E|\Delta t \approx h$, where ΔE is the size of the fluctuation, Δt is its duration, and h is Planck's constant. Feynman has nicely argued that the freedom of the field is explicitly constrained by the interplay of the different terms in the field equation. See R. P. Feynman, "The Qualitative Behavior of Yang-Mills Theory in 2 + 1 Dimensions," *Nuclear Physics B188* (1981): 479–512. (3) The in-effect nonlinear forces leading to the correlations and scaling of fluctuations in a second-order phase transition also lead to the unavailability of so-called irrelevant variables. (4) The constancy of the temperature and pressure in a vessel, whose walls compensate for any deviation, giving equilibrium in thermodynamics and leading to the statistical identity of mean and mode (ergodicity), again determines which fluctuations can show themselves. (5) And the electrical polarizability of an electrostatic shield, making for Gauss's law, prevents our seeing inside details were we to use an external test charge.

25. R. P. Feynman, *Lectures on Gravitation* (Pasadena: California Institute of Technology, 1971), p. 21.

26. To make another point, let me note one other thing walls do: Walls are said to be sharply defined if the insides of one object do not directly affect the insides of another perhaps nearby object. Over very short separations, inside a wall, there is correlation, everything affecting everything else. Over very long distances, through a wall, there is also interaction, the field of one object affecting the behavior of another through that other's external handle. But there is supposedly no direct interaction of the insides of one object with the insides of another. Walls divide up the world into disjoint parts: hiding from each part many of the degrees of freedom of the others, providing only a few external handles for each, and restricting the appearance of most of the internal degrees of freedom – and so creating well-defined objects.

Now much of this will read as if it were a description of the alienation of persons from each other in modern society, which is just what we might expect for the consequences of the division of labor.

27. Massless particles such as photons, moving at the speed of light, cannot be localized statically. There is no Lorentz frame in which they are at rest. But these particles can be seen to be punctiform (and in that sense localized) in their interaction with other particles, as in the Compton effect. They are localized in momentum space (and all particles trade on localizability in position and momentum, by the Heisenberg principle). And radioactive particles are not stable except on a time scale that is short compared to their lifetimes, which is often a good realm in which to work.

28. Again, see Dumont, *From Mandeville*, for a social-intellectual history of this development, as well as on alienation.

29. Frequency (ω) and "crystal momentum" (actually wave vector, $\vec{k}$) are related to velocity, $\partial\omega/\partial\vec{k} = \vec{v}$, acoustically $\omega = c|\vec{k}|$, c being the speed of sound. Those vibrations are in accord with the spatial symmetries of the lattice. Insofar as the lattice is regular and so those symmetries are very well defined, one can attribute a sharp location (in momentum or $\vec{k}$ space) and size to each such kind or particle of vibration. More generally, symmetries and orderliness define particles through their capacity to localize objects, albeit in a peculiar space.

30. Technically, to be stable is to be the ground state (the vacuum) or to be some extent resilient to coupling with

the continuum, and hence in the latter case not quite so readily to make a virtual transition real. Put differently, our time of observation is suitably chosen so that the particle has a well-enough-defined mass. See Anderson, *Basic Notions,* pp. 70–72, for a nice discussion of quasiparticles in a Fermi liquid.

31. In chapter 5 we emphasize that looking is a form of insult, but hopefully a gentle form.

32. T. Clarke, "Seeing Surfaces and Physical Objects," in M. Black, ed., *Philosophy in America* (Ithaca, N.Y.: Cornell University Press, 1965).

33. Technically, as far as I can tell, philosophical and theoretical concerns about the nature of observation within the quantum mechanical realm (as in the Einstein-Podolsky-Rosen paradox) do not alter any of this. Practicing physicists learn to think of particles in *just* the right way so that they are objective in my sense. As in the EPR paradox, the usual difficulty is when the objective individuals are not individual particles but coherent superpositions of particles, and observation destroys that coherence. Then we need a good theory of what observation does (quantum mechanics) and a good model of the state before the observation took place (based on previous observations and a priori expectations). N. D. Mermin, "Bringing Home the Atomic World: Quantum Mysteries for Anybody," *American Journal of Physics* 49 (1981): 940–43, shows just how peculiar are the predictions of quantum mechanics and its mode of summing angular momenta. But quantum mechanics does predict just what we see in observing a spin zero coherent superposition of two spin one-half particles, and so it makes that state objective in that our observation and the imputed original state vector are separated. J. A. Wheeler says that in the quantum mechanical realm "It ain't nothing till I call it," to use an umpire's phrase. (Note that "it" and "I" are separated here.) But again, in practice this does not seem to change what physicists learn to mean by observation or reality or objectivity. For example, coherent states are real for physicists, and so is observation that destroys that coherence. On Wheeler, see J. Bernstein, *Quantum Profiles* (Princeton, N.J.: Princeton University Press, 1991). See also note 35; chapter 3, note 30; chapter 5, notes 24, 32.

34. "Degeneracy" refers to a situation in which two or more different particles (or states) apparently have the same mass (or energy) yet different properties, a situation which physicists find out about because in some situations the particles do behave differently. Namely, physicists expect or find that there is some situation in which the degeneracy "splits" (the Hamiltonian loses its symmetry) – so that particles then exhibit different masses as well as different properties.

As for the muon and the electron, in physics it is a cause for puzzlement and crisis when two objects that otherwise have the same properties do different things (here, possessing different masses), or when what would otherwise appear to be a single object, having a unique mass and lifetime, does more things than its properties allow (as for the *K*-zeros of chapter 3). Note that if two objects have different properties they cannot be said to do the same things; otherwise, we would not know that they had different properties. For properties or names are handles onto an object, and different handles presumably have different consequences. See chapters 3 and 5.

35. At the microscopic level, the same kind of particles are individually indistinguishable, although the roles they play – say being "the atom at position x in the lattice" – are quite distinguishable. We cannot say which of the particles is playing a particular role.

Some of the good degrees of freedom are associated with collections of mutually exchangeable identical particles. Roles have distinct names, but particles exhibit an "exchange symmetry."

36. Here I am thinking of what I understand – in my reading of both S. Kripke, *Naming and Necessity* (Cambridge, Mass.: Harvard University Press, 1980) and the lectures of T. S. Kuhn – to be the nonarbitrariness of names and the embeddedness of the lexicon. See also T. S. Kuhn, *The Road Since Structure, Philosophical Essays 1970–1993* (Chicago: University of Chicago Press, 2002).

37. By the way, the identity of the morning and the evening stars, and of Venus, is not only a philosophical problem but a practical one.

38. Here I am thinking of collective excitations and quasiparticles in metals and in nuclei (Fermi liquids). See Anderson, chap. 3 of *Basic Notions.* Of course, those concerned with interaction energies, like chemists or nuclear physicists, will be less impressed with additivity understood in the ways I have described it (but not either stoichiometrically or if we keep track of mass or energy, for then additivity still applies). And quasiparticles will only add up if we are looking at the margin; otherwise, there is too much interaction for those quasiparticles to remain isolated from each other and so be treated as individuals.

Note that being an individual does not imply distinguishability – see note 35 above.

39. "Parameters that have values in a composite system equal to the sum of the values in each of the [identical] subsystems are called *extensive* parameters." H. B. Callen, *Thermodynamics* (New York: Wiley, 1960), pp. 9, 31. These are also said to be conserved properties since their total value is the same before and after putting the particles together.

We might think of extensive properties as the quantitative properties, in contrast to intensive properties which need be just qualitative. Even if we learn to measure intensive properties numerically, as in a temperature or in color, when the particles are put together the intensive properties are typically averaged rather than summed. (As we shall see in the next note, this is about diffusion as a mechanism for the setting in of equilibrium or of a stable state.) And in many cases, there may well be no way of attributing a well-defined intensive property to the mixed collection.

40. In thermodynamics, intensive parameters are partial derivatives of the fundamental equation $U = U(S,V,N)$, where U, S, V, and N are energy, entropy, volume, and number; their equality before and after putting things together reflects the system's being in equilibrium. Callen, *Thermodynamics,* p. 21. See also note 42 on the Darwin-Fowler method.

Given a sufficiently rough spatial mesh, intensive properties may be attributed to each point in space – they are local. Technically, statistically, an intensive variable, I, is well defined in the infinite volume limit ($N \rightarrow \infty$, density $= N/V$ constant). If $I = I(q(t), p(t))$, then $I = \bar{I} + \mathcal{O}(N^{-1/2})$ if t is large enough. Thermal diffusion, an *averaging* process (think of relaxation as an approximation to the heat equation), eliminates the initial differences in temperature, for example. From another perspective, if the mesh is rough enough, the various points within a particle are statistically independent of each other. When we measure an intensive property, the fluctuations of the property over the particle are in effect damped by the central limit theorem. At phase transitions we do not have this independence. See G. Parisi, *Statistical Field Theory* (Redwood City, Calif.: Addison Wesley, 1988), pp. 2, 8–9.

To be sure, in this section I am pushing an analogy, taking notions natural in thermodynamics and trying to analogize them to elementary particles. I suspect that the notion of an intensive variable may not have an analogy in elementary particles, except when they are considered as thermal and statistical. We might think of how temperature has entered particle physics through models of high-energy systems, whether it be the Big Bang or statistical accounts of the de-excitation of nuclei.

41. Moreover, as we might expect from its definition, entropy is constant (it is conserved) in thermally isolated slow interactions (called adiabatic).

42. Recall from note 40 that thermal diffusion is supposed to eliminate any initial internal variations in a particle if we wait long enough (and if its internal parts are statistically independent of each other). If the parts of an object are otherwise independent, then fluctuations are independent and the central limit theorem applies. This is, roughly, how kinetic theory and statistical mechanics make it clear why fluctuations ($<\Delta E^2> \approx N$) are in general so small proportionately (except at phase transition); namely, why there can be equilibrium, and so why intensive and extensive properties are separable. This separability is also seen in the Darwin-Fowler derivation of the partition function of the canonical ensemble, when the temperature is "pulled out" in the steepest descent derivation. See Schroedinger, *Statistical Thermodyamics,* pp. 31–33. (I am unsure if this analogy of statistical mechanics applies usefully to particle physics.)

Put differently, equilibrium (when forces are zero and potentials are quadratic since the first-order term is zero) allows for a linear response function (from that quadratic potential). And so we have an extensive variable and a co-ordinate intensive variable (the suitable partial derivative) – for example, $\Delta W = -p\Delta V$, where ΔW is mechanical work, p is pressure, and ΔV is the change in volume.

43. See, for example, chapter 3, where we discuss family models of the interaction of particles and the particles' systematic groupings. Of course, the exchanged particles are just what make up the field. But this is, at first glance, just a particulate world. See the next note.

44. Still, it has always been attractive to physicists to give a most parsimonious account, here in the sense of a purely particulate account. See, for example, J. A. Wheeler and R. P. Feynman, "Classical Electrodynamics in Terms of Direct Inter-Particle Action," *Reviews of Modern Physics* 21 (1949): 425–34, and "Interaction with the Absorber as the Mechanism of Radiation," *Reviews of Modern Physics* 17 (1945): 157–81. They formulate the dynamics purely in terms of action at a distance, with no degrees of freedom of its own given to the field. Propagators and Green's functions may be seen to represent action at a distance, or as representing a force that is propagated by an exchanged particle, all interactions occurring at vertices (each vertex can be expanded in higher orders, so it is not quite a point).

Quantum field theory in its perturbative formulation looks very much like a story of the interaction of particles alone (virtual and real) – but it does feature an infinite series of those interactions (or in the case of the scattering matrix formulation, an infinite number of simultaneous equations). In electromagnetism, the first term in the series, one photon exchange, is equivalent to the inverse square law.

45. On "flexible specialization" in manufacturing, see M. Piore and C. Sabel, *The Second Industrial Divide* (New York: Basic Books, 1984).

46. Of course, for situations that are manifestly like fluid flow – such as hydro-

dynamics or a magnetic field – it makes even less sense to restrict one's description to walls and particles. See Buchwald, *From Maxwell,* on the Maxwellians and their field orientation, where charge was distributed as well. Moreover, it might be noted that phenomenological space itself (space as we experience it and measure it) is a consequence of the gravitational field's provisioning the universe with space's metric tensor, so to speak.

47. Fields are said to be local if their value at a point depends on nearby values of the field; the field is expressed as differential equations (for which see chapter 2). See J. D. Bjorken and S. Drell, *Relativistic Quantum Fields* (New York: McGraw Hill, 1965), pp. 3–5. And, as we discuss, interactions can also be local, the interaction of a particle with a field being in terms of the particle's properties and the value of the field at the particle's location. If we actually observe a field by employing a test particle to feel it out, then that almost seems to require a local interaction. Note that if there are quantum fluctuations or statistical fluctuations, then continuity (and dependence, as in chapter 2) does not manifestly apply and the infinitude of degrees of freedom is not obviously constrained.

Kenneth G. Wilson, in "The Renormalization Group: Critical Phenomena and the Kondo Problem," *Reviews of Modern Physics* 47 (October 1975): 773–840, points out that differential equations and partial differential equations are a consequence of the existence of continuum limits and "statistical continuum" limits:

> One of the most basic themes in theoretical physics is the idea that nature is described locally. The basic equations of all physics are local. For example, Maxwell's equations specify the behavior of electric and magnetic fields in an infinitesimal neighborhood of a point x. In order to be able to specify local equations it is necessary to define continuum limits, namely the limits which define derivatives. The idea of the derivative and the idea of a continuum limit that underlies the derivative is therefore of great importance in all of physics.
>
> It is now becoming clear that there is a second form of continuum limit, called the statistical continuum limit, which also has a very broad range of applicability throughout particle physics. In the statistical continuum limit functions of a continuous variable are themselves *independent* variables. For example, the electric and magnetic fields throughout space can be the independent variables in a statistical continuum limit. This happens in statistical or quantum mechanical problems where there are field fluctuations, so that one has to compute averages of an ensemble of fields. . . .
>
> There are two ways in which a statistical continuum limit can arise. The obvious way is when the independent field variables are defined on a continuous space; the case of statistical or quantum fluctuations of the electromagnetic field is an example. . . .
>
> The second source of statistical continuum limits is the situation where one has a lattice with a fixed lattice spacing, usually an atomic lattice. The number of independent variables (i.e. independent degrees of freedom) at each lattice site is fixed and finite. The continuum limit arises when one considers large size regions containing very many lattice sites. When the lattice is viewed on a macroscopic scale one normally expects the lattice structure to be invisible. That is, large scale effects should be describable by a continuum picture making no reference to the lattice spacing. (pp. 773–74)

48. Here I have precluded consideration of turbulence. And shock waves and caustics are here treated as special if extraordinarily important and informative cases.

49. I should note that in actual water flow we might imagine that two spigots could affect each other's flow at the respective sources. This is not the case for electrodynamics or gravity, at least as conventionally presented. But when electrodynamics is in nonlinear regimes (fields being large and bulk matter being present, or in vacuum polarization) there are canonical ways of accounting for such nonlinearity yet preserving or at least protecting the ideal of linear superposition (the nonlinearity being attached to the material). See also note 54. See J. D. Jackson, *Classical Electrodynamics* (New York: John Wiley, 1975), pp. 10–13.

50. The term comes from C. W. Misner, K. S. Thorne, and J. A. Wheeler, *Gravitation* (San Francisco: Freeman, 1973), p. 367.

51. Rather than thinking that the Aharonov-Bohm effect proves the reality of the electromagnetic vector potential, we might say that the vector potential, A_μ, assures us, especially in just this effect, of an objective world.

52. Following Misner, *Gravitation,* pp. 370–71, and to some extent repeating note 20, and being a bit speculative:

One wants to automatically link particle properties ($\vec{J}$ or current) to the field ($\vec{A}$ or vector potential) so that the particles are the source of the field $\vec{A} = \vec{J}/r$) and so there is superposition; to link field properties ($\vec{A}$) to the walls ($\vec{E}$ and $\vec{B}$, the electric and magnetic fields respectively) – so getting path independence and objectivity (Stokes law, that is, curl $(\vec{E} - \partial\vec{A}/\partial t) = 0$, where I take the conventional fields to represent the walls because of Gauss's law); and to link those walls back to the particle properties so as to get shielding (Gauss's law, div $\vec{E} = J_4$).

53. Misner, *Gravitation,* pp. 367–68. In Wheeler, *A Journey,* he speaks of how mass and spacetime are in the "grips" of each other.

54. Here I am thinking of classical hydrodynamics. Another example: Quantum field theory defines particles and their fields at the same time. The requirements that fields be smooth, connected to their particulate sources, conservative, and linearly additive (superposition), and that they take into account path dependence, are powerful constraints on quantum field theories (or classical Maxwellian electromagnetism, for that matter). But those requirements have come to be seen as necessary if the field is to do what a field must do. I should note that physicists often think of fields as independent of their sources – one of the main practical justifications for the notion of a field. Different sources may produce the same field at a particular point and so the same force on a test particle – and so physicists justify their giving a field its own integrity. Moreover, such fields, like flows of water, can be shown to carry momentum and energy. See Jackson, *Classical Electrodynamics,* p. 3.

55. W. B. Arthur, "Competing Technologies, Increasing Returns, and Lock-In by Historical Events," *Economic Journal* 99 (March 1989): 116–31, and "Positive Feedbacks in the Economy," *Scientific American* (February 1990): 92–99; and C. Sabel, *Work and Politics* (Cambridge: Cambridge University Press, 1982).

2. TAKING APART AND PUTTING TOGETHER

1. The usual question concerning the relationship of parts to wholes is almost always a philosophical question, unless the problem is to figure out a mechanism or an interaction that does the required work. By the way, I have not discussed in this chapter many other mechanical partitions, such as those that do not obviously depend on interdigitation. See chapter 4. The notion of mechanism is here used quite generically. See E. J. Dijksterhuis,

The Mechanization of the World Picture (Oxford: Clarendon, 1961), pp. 495–501.

2. David Hume, part 2 of *Dialogues Concerning Natural Religion* (New York: Penguin, 1990), p. 53.

3. B. Bettelheim, *A Good Enough Parent* (New York: Knopf, 1987), pp. 172–73.

4. S. Weinberg, "Why the Renormalization Group is a Good Thing," in A. Guth, K. Huang, and R. Jaffe, eds., *Asymptotic Realms of Physics* (Cambridge, Mass.: MIT Press, 1983), p. 16.

5. On Nature being referred to as "her," see E. F. Keller, *Reflections on Gender and Science* (New Haven, Conn.: Yale University Press, 1985).

6. More precisely, the angle variable is the frequency of rotation rather than the period. These action-angle variables, which are constants of the motion, for a single planet system, lead to a natural correspondence with quantum mechanical accounts of an electron "orbiting" a nucleus. They also allow for a natural theory of the perturbing effects on a planet's orbit due to other planets in the system.

7. In *The Structure of Scientific Revolutions* (Chicago: University of Chicago Press, 1970), T. S. Kuhn has emphasized how scientists choose their cases – namely, the choice of paradigmatic exemplars of actual problem solving.

It is often the case that the analytically discovered degrees of freedom – the parts – only do the compositional work they have to do when we allow for their doing it approximately and "in the limit": when temperatures are very high or low, as in a gas or solid; when the number of composing elements or parts is very large, as in statistical mechanics; or when the composing elements are infinitesimally small, as in the calculus. What is interesting is that physicists are willing to allow for such a regime of approximation – close enough being good enough. They seem to be willing to go very far in order to be able to keep their commitment to analysis and composition. Of course, there are many real world situations that correspond to these limits.

8. See chapters 1 and 2 of M. H. Krieger, *Marginalism and Discontinuity: Tools for the Crafts of Knowledge and Decision* (New York: Russell Sage Foundation, 1989) on adding up the world and on marginalism more generally. Some of my text here is an excerpted and revised version of what appears there. The distinction between marginalism and discontinuity is surely not absolute. Marginal parts, as in the periodic table of the elements, could be treated as components in a chemical reaction, for example.

9. Not only is the whole different from the parts, but the whole is a matter of having all the parts just in place. Now, keystoning is a risky and precarious mode of composition. A coded message that makes no sense unless we can decode it completely is quite safe, but if there are errors in transmission of the coded message, it may not be decipherable at all. In general, it is more robust to have a hierarchical composition, one with stable whole subassemblies which then fit together, especially if there is some redundancy and backup as well.

Note that chemical reactions may be thought of as a plenitudinous set of random collisions of molecules, the kind and quantity of the resulting complex molecules an artifact of the (free) energetics.

The techniques employed for constructing general equilibria of a market system, given its components, involve a particular method of putting things together: an actual process of haggling, *tâtonnement,* as a means of successive approximation or mathematical devices to achieve the same result. See H. Scarf and T. Hansen, chap. 1 of *The Computation of Economic Equilibria* (New Haven, Conn.: Yale University Press, 1973).

10. For differential equations, this is a matter of the separation of variables, namely, a separate differential equation for each variable alone.

11. In chapters 1 and 4 we discuss how that hiding is rarely perfect, and there are consequences to those fluctuations.

In different regimes, the consequences of the same marginal additions (of energy, atoms, . . .) may well be very different. An extra bit of accreted mass has different consequences for a Sun-like star than for an almost black hole.

12. More generally, stories of origin (as in Genesis) have prescribed direction and sequence. So stories of marginal addition often seem almost irretrievably historical, the path they follow prescribed and unavoidable. Yet, in fact, it turns out that one of the desires we have is to avoid that historicity; and, as we shall see, it is often the task of additive strategies such as the calculus to tell a story of marginal changes in which there is path independence.

13. Sometimes (see chapter 4) those discontinuous changes can be smoothed out, and we can watch how new degrees of freedom appear. "But sometimes the choice of appropriate degrees of freedom is not just a question of large or small wavelength, but a question of what kind of excitation we ought to consider. At high energy the relevant particles are quarks and gluons. At low energy they're massive pions. What we need is a version of the renormalization group in which as you go from very high energy down to low energy you gradually turn on the pion as a collective degree of freedom, and turn off the high-energy quarks" (Weinberg, "Why the Renormalization Group is a Good Thing," p. 17).

14. See, for example, M. Kac, *Statistical Independence in Probability, Analysis and Number Theory* (Washington, D.C.: Mathematical Association of America, 1959).

15. The interesting question is how far out in the tails of the distribution do you still believe that you have a Gaussian, that is, statistical independence (at least as much as is needed for the central limit theorem to apply) rather than bias. More technically, the theory of large deviations offers corrections to the central limit theorem so that what might at first seem like a deviation from independence is in fact affirmed as confirmation of it. See, for example, R. Arratia and L. Gordon, "Tutorial on Large Deviations for the Binomial Distribution," *Bulletin of Mathematical Biology* 51 (1989): 125–31.

J. Ford, "How Random is a Coin Toss," *Physics Today* (April 1983), discusses how deterministic processes can yield random outcomes, by employing notions of chaos and the acute sensitivity of such dynamical systems to initial conditions. See also P. Diaconis, S. Holmes, and S. Montgomery, "Dynamical Bias in the Coin Toss," SIAM *Journal* 49 (2000): 211–35.

16. See chapter 3 note 19.

Then, in dialectical response, we may try to show how suitably chosen parts, more or less independently cumulated, give us what we might otherwise take to be the historical and designed world. More generally, we try to find dumb parts which, when composed, give us the actual historical and patterned world. As we shall see, those dumb parts need not be independent. They may depend on each other, or be additive, or be procedures. On the Argument from Design and the dialectic with dumbness, see M. H. Krieger, "Temptations of Design: A Meditation on Practice," *Research in Philosophy and Technology* 10 (JAI Press, ed. F. Ferré, 1990): 217–30, reprinted as part of "The Manufacture of the Sacred, the Reenactment of Transcendence, and the Temptations of Design," in M. H. Krieger, *What's Wrong With Plastic Trees? Artifice and Authentic-*

ity in Design (Westport, Conn.: Praeger, 2000), pp. 75–94.

17. See M. H. Krieger, "Theorems as Meaningful Cultural Artifacts: Making the World Additive," *Synthese* 88 (1991): 135–54.

18. In 2011 the first part of this paragraph appears quite naive.

Technically, we have what is called a martingale (the expected future price is just the current price). All information about the future and the past is summarized in the current price, and hence the next step is unpredictable. The critical application of this observation to the pricing of options is F. Black and M. Scholes, "The Pricing of Options and Corporate Liabilities," *Journal of Political Economy* 81 (1973): 637–54. For an interesting and radical argument, in which it is suggested that trading depends on noise and not on information, so that randomness is rather more deeply influential on actual prices, see "Noise," in F. Black, *Business Cycles and Equilibrium* (New York: Blackwell, 1987), pp. 152–72. The description here is to say the least abstract; the bidders are too dumb. I have left out differences in attitudes toward risk, the fact that information is available through the bidding process, and that individuals try very hard to have and would prefer to have some information asymmetry, of course in their direction. But if this is not the case, then option-pricing schemes (which depend on the random-walk character of prices) allow one to look for arbitrage opportunities as a way to make money.

19. Several points: Physicists would say "configuration" rather than "arrangement." Collisions in a gas at room temperature and pressure occur very many times a second for each particle. Quantum mechanical considerations do not change what I say in the text. And proofs of the in-effect independence of the states (ergodicity) are a small industry in itself.

20. Here I have deliberately chosen to list extensive properties, their additivity being one of the right features. Note that independence of parts, here parts of a system, is a sufficient condition for there to be an additive extensive property, such as energy. For, by independence, $P_A P_B = P_{AB}$, where P_A is a probability or a density measure; then the logarithms of the P's will add; and those $\log P_A$'s will be extensive variables (say proportional to energy). See L. Landau and E. Lifshitz, secs. 1, 2, and 4, in pt. 1 of *Statistical Physics*, 3rd ed. (Oxford: Pergamon, 1980).

21. The arrangements or states are represented by the equilibrium distribution of energies, and almost all are at the average energy.

Here, the internal degrees of freedom are "irrelevant." In nineteenth-century terms, we might think of a very general law of errors (or what we call a central limit theorem) that would say that sums of "randomish" variables are in general "Gaussianish" – namely, the sums have nice means and variances, their comparatively small higher moments then hiding most of the degrees of freedom. On "generalized rigidity" (and infinite volume limits), which might be taken to be a consequence of such a generic central limit theorem, see P. W. Anderson, chap. 2 of *Basic Notions of Condensed Matter Physics* (Menlo Park, Calif.: Benjamin/Cummings, 1984). See also chapter 1, note 13. See T. Porter, *The Rise of Statistical Thinking* (Princeton, N.J.: Princeton University Press, 1986), for a history of these ideas in a societal context.

At points of phase transition independence breaks down (technically, an ergodic to nonergodic transition), and there is pervasive correlation among arrangements, and so fluctuations can be unbounded.

22. As we shall see in chapter 3, rather than starting out with parts which we

then take as independent, we can say that any arrangement of parts that is not forbidden must happen ("plenitude," a sort of ergodic theorem), and if there is a very large number of parts, we again achieve many of the consequences of independence. Independence and plenitude, as they define the parts, are a curious and radical form of the alienation and methodological individualism characteristic of liberalism. See Porter, *Rise of Statistical Thinking,* about the relationship of statistics, alienation, and nineteenth-century liberal society.

23. A work of aleatory art is seen as whole and coherent either because we impute dependence to it or because local randomness need not lead to what appears to be global randomness.

24. The commitment could be to difference equations and to spreadsheets. For a straightforward discussion of locality and its relationship to special relativity, see J. D. Bjorken and S. D. Drell, *Relativistic Quantum Fields* (New York: McGraw-Hill, 1965), pp. 3–5. See chapter 1, note 47.

25. See the quotation by Feynman in chapter 1, at note 25, which indicates how one sets things up experimentally so that the separation I refer to is nicely exhibited.

26. In the first case, points that are on different horizontal lines are not at all nearby; in the second, diagonal points are not so nearby (but on an L-shaped path to each other); and in the third, points have horizontal, vertical, and diagonal neighbors.

Given local and uniform rules, the consequences of composition are rather different depending on the degree of connectivity. With a merely (horizontal) linear connection, there will never be the high coordination of atoms characteristic of freezing, for example. With greater connectivity among the particles or objects – say planar – the orderliness of freezing does become possible. However, the nature of the phase transition will change depending on the amount and kind of connectivity.

Consider the following simple rule: Connect – with probability c – each point of a square planar Cartesian grid to any nearby point, by a horizontal (I) or vertical (II) or diagonal (III) line. We might ask what is the likelihood of there being a solid or reliable path across the grid. In case I, there is no such path unless $c = 1$. In case I + II, if there is to be a path, c has to be greater than 0.5; and in case I + II + III, c can be somewhat smaller. We have here an analysis of the reliability of a telephone system, or the likelihood of epidemic spread of a disease: whether a call will get through, or the disease will become pandemic. See D. Stauffer, *Introduction to Percolation Theory* (London: Taylor and Francis, 1985).

Recently, the conformal invariance of percolation and the Ising model, at the critical point, has been proven. W. Werner, "The Work of Stanislaw Smirnov," *Notices of the American Mathematical Society* 58 (March 2011): 462–64.

Note how Hamiltonian formulations of dynamics place the degrees of freedom of position and momentum, the q's and p's, close to each other. In so doing, their phase space nicely allows for the dynamical fact – namely, the physics! – that the Hamiltonian flow in that space is very simple: the tangent $\partial H/\partial p$ is equal to dq/dt, for example.

27. On the commitment to continuity (and smoothness, although these are not the same, that deepest realization being a legacy of nineteenth-century mathematical rigor), see the preface, introduction, and chapters 1 and 2 of Krieger, *Marginalism and Discontinuity.* Of course, differential equations can exhibit discontinuities, whether they be due to boundary condi-

tions as in the Schroedinger equation (discontinuity not in the wave function, but discreteness in the eigenvalue spectrum), or caustics, or bifurcations.

The insistence on continuity and smoothness also further restricts the uniform and local parts, for they then have to fit into a smooth manifold. Those further restrictions are nontrivial, for there are putative solutions to those equations, especially if we are doing numerical approximations, which are not at all "physical," to use the term of art. Put differently, in spreadsheet terms, the cell formulas we have been talking about may be "circular." And it is not clear that we shall have a stable or convergent solution no matter how often we recalculate. It turns out that we can often have such unique stable solutions, but we must specify the mode of recalculation and the generic shape of the solution (as we must more generally for methods of numerical approximation).

28. If the parts were the value of a field at each point, we might get hold of these parts with test particles, designed to measure the field point by point without affecting them much. The strategy of probes described in chapter 5 depends on having parts that may be gently probed individually by test particles, for example. As we shall argue there, ontology and observation are intimately connected. More generally, on problems of the existence of field values and of "statistical continuum limits," see K. G. Wilson, "The Renormalization Group: Critical Phenomena and the Kondo Problem," *Reviews of Modern Physics* 47 (October 1975): 773–840. See also chapter 1, note 47.

I suspect it might be argued that dependence that is local and uniform ensures that each part can be considered as an individual. Physically, a test particle can do its work of measuring a field only if interactions are uniform and local (so that we are measuring the same thing, at a point) – and in so doing, the test particle affirms the "individuality" of the field at that point. (A test particle ought not be concerned about what it takes hold of or where it is.)

More generally, parts are often designed to be sufficiently loosely coupled to each other so that their dependence or coupling could be considered quite weak, at least in principle, in some regime. If there are normal modes of vibration, these normal modes are again very good, nicely isolated, degrees of freedom – individuals par excellence, dependence now out of the question except as anharmonicities.

Politically, it might be argued that democracy and fairness, corresponding to locality and uniformity, are the foundation for individualism. Similar arguments could be made in the economic realm.

29. For markets, the instantiation is in the form of the *tâtonnement* or haggling process for the solution of a system of simultaneous equations (balancing supply and demand, in general equilibrium). See Scarf and Hansen, *Computation of Economic Equilibria*.

30. For an account of this irony, presented as physics, see, for example, A. Pais, "The Reality of Molecules," chap. 5 in *Subtle is the Lord . . . : The Science and Life of Albert Einstein* (New York: Oxford University Press, 1982), pp. 79–107.

What is of interest to physicists is not the conservation law. (The divergence of a current is proportional to the time rate of change of a density, $\partial/\partial x\ (D\ \partial n/\partial x) = D\ \partial^2 n/\partial x^2 = \partial n/\partial t$, where n is a measure of density of matter or of probability and D is the diffusion constant, here presumed spatially invariant.) That is a given, largely from the assumption of steady-state behavior or that micro-fluctuations are unimportant. Rather, what was exciting was the derivation of the macroscopic diffusion constant, D, in terms of microscopic variables.

Mathematicians say that martingales are related to harmonic functions. That random walks and diffusion and potential theory are systematically related was not at all obvious. Technically, see M. Kac, "On Some Connections between Probability Theory and Differential and Integral Equations," in K. Baclawski and M. D. Donsker, eds., *Mark Kac: Probability, Number Theory, and Statistical Physics* (Cambridge, Mass.: MIT Press, 1979), pp. 322–50.

31. The difference between the strategy of differential equations (dependence) and that of the integral and differential calculus (additivity), as I describe them here, is that in the first strategy the crucial idea is continuity, and in the second it is arithmetic. The fundamental theorem of the calculus shows how these are intimately connected: given continuity, there is a good arithmetic.

32. On path invariance, see "Adding Up the World," chap. 2 in Krieger, *Marginalism and Discontinuity,* on the calculus and its culture.

33. If there are many parts, one resorts to combinatorial devices for doing the counting, or perhaps one finds an approximation to the sum. For much of what follows, and in greater detail, see Krieger, "Adding Up the World." See chap. 6.

34. Recall that if there is independence then there can be an additive extensive property, since the logarithm of a product is a sum. See note 20. But the parts have to have very weak mutual interaction energies if there is to be independence, which is not the case when there is a chemical reaction.

35. Or, to choose a somewhat different situation, we decompose a differential equation into its independent components ("separation of variables") according to the symmetries of a situation, to again get "linear" additivity of eigenvalues along each axis. Recall note 10 and its associated text.

36. On further strategies for addition, see Krieger, "Theorems as Meaningful Cultural Artifacts."

37. By providing a good definition of limits, whether as tangent vectors to a line (velocities), or in terms of Riemann rectangles and sums for figuring out an area, the summed length or area turns out to be invariant to how you segment a path or line, or how you divide up an area. As a consequence of continuity and the provision of canonical definitions of limits, the fundamental theorem of the calculus says that sums are invertible (derivatives and integrals are inverses of each other), and small changes have small (that is, marginal or differential) effects on the sum. The calculus provides for a linear world (through tangent vectors and Riemann rectangles), one that is additive – even if the objects to be added are changing and curved. (More sophisticated methods, such as Lebesgue integration, allow for not quite so smooth a world.)

38. A. Marshall, the preface in *Principles of Economics* (1890) (London: Macmillan, 1920). Jevons is perhaps the crucial figure here. See M. Schabas, "Alfred Marshall, W. Stanley Jevons, and the Mathematization of Economics," *Isis* 80 (1989), pp. 60–73.

39. See F. P. Brooks, "No Silver Bullet: Essence and Accidents in Software Engineering," *Computer* (April 1987): 10–19, for an exposition of the practical issues involved in actual work within a productive bureaucracy – from the computational point of view.

I have deliberately not used systems analysis here, largely because I want to stay within the realm of the very concrete models. Also, I want to avoid the language of efficiency and emphasize a language of abstraction (disjointness and interpretability). Chapters 1 and 3 might well be reconceived in systems-analytic terms. Herbert Simon's work, from studies of bureaucracy to studies in artificial intelligence, epitomizes the connection I discuss here.

40. J. March has emphasized this point in his garbage-can model of organizations. J. G. March and J. P. Olsen, in chap. 2 of *Ambiguity and Choice in Organizations* (Bergen, Norway: Universitetsforleget, 1976).

41. A clockworks may also be conceived in terms of an articulated mechanism. Gears, pulleys, and levers link the various sub-mechanisms, which sub-mechanisms are made up of gears, pulleys, and levers themselves. What flows through such a system is energy and momentum (and information). Can one design a universal set of mechanisms which, through their linkages, are able to do everything? That is one of the great traditional questions of engineering and of mechanical philosophy.

42. Here, I emphasize that "objects" or data or applicants are perhaps the central feature of object-oriented programming, and that bureaus or "methods" or procedures are secondary. See, for example, B. J. Cox, *Object Oriented Programming* (Reading, Mass.: Addison Wesley, 1988). But, in my conception of object-oriented programming, drawn from simulation languages like GPSS, there is no master sending messages to objects except perhaps for a clock. Here objects seem to trigger themselves, so to speak, either by threshold variables or by internal clocks which act as threshold variables.

43. I say "so far" in the text because the computational model is beginning to have deep influence within theoretical physics, whether it be cellular automata or special-purpose computers to do gauge-field theory calculations. See, for example, S. Wolfram, ed., *Theory and Application of Cellular Automata* (Singapore: World Scientific, 1986).

The bureaucratic terms become those of systems, queuing, and network theories, providing new kinds of degrees of freedom. What we would now call systems-analytic war research on microwave devices by J. Schwinger, treating devices in terms of inputs, transfer functions, and outputs, led to his formulation of quantum electrodynamics in terms of the black boxes of an S-matrix, in which scattering interactions are expressed in terms of inputs and outputs. See J . Schwinger and D. S. Saxon, *Discontinuities in Waveguides* (New York: Gordon and Breach, 1968), p. ix. For more in this vein, see S. S. Schweber, "The Empiricist Temper Regnant: Theoretical Physics in the United States 1920–1950," *Historical Studies in the Physical Sciences* 17 (1986): 55–98.

In fact, my claim in the text about the limited influence of bureaucracy is perhaps belied by perturbation theory taken as an interpreted disjoint hierarchy, as well as by S-matrix theory.

44. Such comparisons may require that we do many such simulations, since fluctuations and nonequilibrium runs need not be uncommon when the number of simulations runs is small. But perhaps those fluctuations (rather than averages) are just what Nature exhibits.

45. See, for example, P. Galison, *How Experiments End* (Chicago: University of Chicago Press, 1988), on "cuts" in the data.

46. In computational terms, eventually – again in actual practice – we have to write some lines of code that calculate and do work. The practical question is how primitive and simple these ought to be.

47. The Hamilton-Jacobi equation tells, for example, the next step in terms of propagated wavefronts. Consider as well the process of haggling or *tâtonnement* in general equilibrium economic theory as a way of finding equilibrium, the optimum state, for which see Scarf and Hansen,

Computation of Economic Equilibria. On simulated annealing, see S. Kirkpatrick, C. D. Gellatt, Jr., and M. P. Vecchi, "Optimization by Simulated Annealing," *Science* 220 (13 May 1983): 671–80.

48. H. Abelson and G. Sussman, *Structure and Interpretation of Computer Programs* (New York: McGraw Hill, 1985), p. xiii. It should be noted that one of LISP's central features is "procedural abstraction," namely, the details of how a procedure actually does what it does are of lesser importance than what it does.

49. I should note, again and as for bureaucratic analysis, comparatively little actual physics yet uses this mode of analysis and composition. But there is currently much talk about using computational models to understand physically interesting situations.

50. "In fact, [in integration by parts] if the integrated part does not disappear, you restate the principle, adding conditions to make sure it does!" R. P. Feynman, R. B. Leighton, and M. Sands, vol. 2, lecture 19, of *The Feynman Lectures on Physics* (Reading Mass.: Addison-Wesley, 1964), p. 5.

Andy Pickering has written extensively about the relationship of artisanal skills, problem choice, and comparative advantage. See, for example, *Constructing Quarks* (Chicago: University of Chicago Press, 1984).

3. FREEDOM AND NECESSITY

1. I. I. Rabi, as quoted frequently in many references. I have found no authoritative source.

2. B. Spinoza, pt. 1 in *Ethics* (1677). I might well have quoted Hegel. But in either case there is an irony. For the issue in this chapter is not causality and free will; rather, it is the relationship of possibility per se to necessity – and here what is possible is necessary.

3. Technically, degeneracy usually refers to having the same mass or energy but different properties. The Hamiltonian is unchanged by a symmetry, one which is not yet broken (say by an external field). We would want to say that the electron and the muon ought to be degenerate, by the identity of their properties, but they are manifestly not, having very different masses. And so we look for other properties and associated interactions that then account for that difference. See also chapter 1, note 34.

4. These three distinct families of elementary particles, each headed by a different neutrino and its associated charged lepton such as the electron, are distinguished by their "flavor" (and the flavors of their associated quark doublets, *u* and *d* for the electron-neutrino, for example), a new degree of freedom. The difference between the electron and the muon is now associated with a great many other differences. The muon is necessary in that it belongs within the muon-neutrino family, and that family belongs within an even larger context of families.

5. Much of this section comes from M. H. Krieger, "The Elementary Structures of Particles," *Social Studies of Science* 17 (1987): 749–52. That title roughly mirrors C. Lévi-Strauss's *The Elementary Structures of Kinship (Les structures élémentaires de la parenté,* 1949) (Boston: Beacon Press, 1969). As for exchange, the crucial work in this context is M. Mauss, *The Gift* (1924; reprint, New York: Norton, 1967), who emphasizes the importance of "unequal" exchanges, namely, implicit and explicit future obligations. The anthropologists have the economic analogies in mind when they describe these kinship systems. It should be noted that Lévi-Strauss is perhaps ambiguous about whether he is concerned with prescriptive or proscriptive ("closed" or "open") systems, marriage and incest rules, respectively, which are identical only under the assumption of plenitude. That ambiguity is, I gather,

of great importance in practical anthropological work. For physicists, plenitude is assumed to apply, and so there is no difference. For more on such exchanges, see "Sticky Systems" in M. H. Krieger, *Marginalism and Discontinuity: Tools for the Crafts of Knowledge and Decision* (New York: Russell Sage Foundation, 1989).

6. A. O. Lovejoy, *The Great Chain of Being* (Cambridge, Mass.: Harvard University Press, 1936), p. 52. See also, on plenitude in a physical science context, H. Kragh, *Dirac* (Cambridge: Cambridge University Press, 1990), pp. 270–74.

7. The interaction may be said to happen even if energy is not conserved, "virtually," in quantum mechanical terms. And those virtual interactions do have effects – but we never see those interactions directly. See chapters 1 and 4 on fluctuations.

8. Conversely, "Long-lived theoretical entities, which don't end up being manipulated, commonly turn out to be wonderful mistakes." I. Hacking, *Representing and Intervening* (Cambridge: Cambridge University Press, 1983), p. 275. On my use of "fetish," see "Economy and Perversion," in Krieger, *Marginalism and Discontinuity.*

9. Here I am referring to phase space considerations and to statistical weight ($e^{-E/kT}$) effects.

10. Since we have a mostly electrically neutral everyday world (but think of lightning and thunder), we do not directly see the strong electrical forces. Conversely, everyday bar magnets reflect the unshielded un-neutralized effects of many atoms, and hence magnetic forces appear quite strong to us. Qualitatively, electrical forces are inverse square, while magnetic forces are dipolar and so inverse fourth power. See chapter 4, note 16, on why the proton is used for comparison.

11. Here temperature is the energy variable. Note that the critical point, expressed in terms of temperature and pressure, is a function of the atomic interaction strength.

12. In actual theories of the sort I describe, the observed phenomena often depend on squares of these amplitudes, so that anti-cancellation leads to four times that of a single contribution (that is, 2^2), not twice.

13. In the case of light, each history may be said to have attached to it a phase factor, $e^{i\theta}$ – a vector of unit length pointing in a certain direction – the direction depending on that particular history. For an elementary exposition, see R. P. Feynman, *QED* (Princeton, N.J.: Princeton University Press, 1985). I cannot find the word "democracy" in Feynman's work. See note 17, below. His teacher, Wheeler, uses it to describe Feynman's work in C. Misner, K. Thorne, and J. A. Wheeler, *Gravitation* (San Francisco: Freeman, 1973), pp. 418–19, 499–500.

14. Of course, a great deal of work is required to set up equilibrium. See the Feynman quote in chapter 1, at note 25. Moreover, Feynman argues in *Lectures on Gravitation* (see note 23 below) that the everyday world (not the physicist's laboratory setup) is often the converse of plenitude.

15. Here I am thinking of the Darwin-Fowler derivation of the canonical ensemble ($e^{-E/kT}$). The existence of a temperature is equivalent to the existence of a pole in the steepest descent derivation, which is equivalent to a most popular distribution (states belonging to other distributions literally canceling themselves out, as in interference) – namely equilibrium. See also the next note. See, for example, K. Huang, *Statistical Mechanics* (New York: Wiley, 1963), pp. 206–13; or E. Schroedinger, chap. 6 in *Statistical Thermodynamics* (Cambridge: Cambridge University Press, 1964). Note that plenitude plays its role here in our assuming that all dis-

tributions are equally likely, given the constraints.

16. See chapter 2, where independent parts were again described in these terms of statistically independent arrangements or configurations (points in phase space). We are in the realm of statistical mechanics and its ensembles of independent states. See the previous note for a justification of the weight factor.

17. See, for example, R. P. Feynman, *Statistical Mechanics* (Reading, Mass.: Benjamin/Cummings, 1972), where the notion of democracy is applied to statistical mechanics, and where the formal analogy to quantum mechanics and path integrals as sums over all paths is explicit. On ergodicity, see, for example, G. Parisi, chap. 18 on "The Approach to Equilibrium," in *Statistical Field Theory* (Redwood City, Calif.: Addison Wesley, 1988).

18. If the number of states goes as E^D, where E is the total energy and D is the number of degrees of freedom, and given the weight of $e^{-E/kT}$, then the contribution to the sum of weights is greatest for the term when $E = DkT$ – that is, $(DkT)^D e^{-DkT/kT}$ is the maximal term – as we would have expected from the equipartition theorem, for which see chapter 4, note 54 (where k is appropriately factored out).

19. In this sense, the Invisible Hand is the opposite of the Great Chain of Being, with the latter's continuity from one scale to the next.

20. Or so it is alleged in accounts of general equilibrium. See, for example, R. Dorfman, P. Samuelson, and R. Solow, *Linear Programming and Economic Analysis* (New York: Dover, 1958, 1986). Again, the anthropologists knew of market models when they confected their kinship models.

21. See, for example, A. O. Hirschman, *The Passions and the Interests* (Princeton, N.J.: Princeton University Press, 1977); and L. Dumont, *From Mandeville to Marx* (1977; reprint, Chicago: University of Chicago Press, 1983).

22. Fields may be the all-encompassing notions, particles and their properties and exchanges playing no central role, as in J. Z. Buchwald's description of the Maxwellians in *From Maxwell to Microphysics* (Chicago: University of Chicago Press, 1985). Conversely, in modern quantum field theory a field is made up of an infinitude of particles and their production, annihilation, and exchange.

23. Were *we* to look at the world, plenitude seems to be not at all true (although to the medieval mind of the Great Chain of Being it was quite apparent). "Often one postulates that *a priori,* all states are equally probable. This is not true in the world as we see it. . . . There are in the world people – not physicists – such as geologists, astronomers, historians, biologists, who are willing to state high odds that when we look into an as yet unobserved region of the universe, we shall find certain organization which is not predicted by the physics [of plenitude] we profess to believe. . . . How then does thermodynamics work if its postulates are misleading? The trick is that we have always arranged things so that we do not do experiments on things as we find them, but only after we have thrown out precisely all those situations which would lead to undesirable orderings" (R. P. Feynman, *Lectures on Gravitation* [Pasadena: California Institute of Technology, 1971], pp. 18, 21).

24. To use Thomas Kuhn's terms, we have a paradigm that presents us with puzzles. See *The Structure of Scientific Revolutions* (Chicago: University of Chicago Press, 1970).

25. Calculating the amplitude numbers for each process, each fully labeled history, is no mean task, perhaps requiring the solution of the field equations or the Schroedinger equation. To figure

out what we might actually see (namely, an only partially labeled situation), we square the sum of the amplitudes of the indistinguishable processes to obtain measures of the probability distribution of the outcomes.

26. For example, if a spin zero particle decays into a two spin one-half particles, then two combinations are possible: up-down and down-up. If we examine one of the decay products we know about the direction of the spin of the other. See N. D. Mermin, "Bringing Home the Atomic World: Quantum Mysteries for Anybody," *American Journal of Physics* 49 (1981): 940–43, for an exposition of the seeming paradox here.

27. These details include that you also get pairs of pi-zeros as well as pairs of charged pions, associated production of *K*-zero and lambda particles (and the strangeness quantum number), parity violation, that the *K*-zero is not a good degree of freedom of weak interactions (that is, having a well-defined radioactive lifetime, hence the two lifetimes discussed in the text) while the *K*-zero-one and *K*-zero-two are, and are states with almost well-defined CP, etc. Moreover, in 1964 it was discovered that *K*-zero-twos also decay into two pions, but quite rarely, and I also ignore this wonderful complication of CP violation here. See A. Pais, chap. 21 in *Inward Bound* (New York: Oxford, 1986), for a nice survey of *K*-zero physics by a participant.

My example here is confected for expository device. The canonical regeneration example is a pure *K*-zero beam that downstream produces hyperons in matter by *K*-zero-bar plus proton yields lambda-zero plus pi-plus. As for regeneration, the *K*-zero-bars arise from the decay of the *K*-zero-one component of the original beam.

28. The amplitudes are the same for particle and antiparticle. Under the particle-antiparticle transformation, the *K*-zero and the K-zero-bar interchange and the pions (pi-plus and pi-minus) also interchange.

29. S. Weinberg says that a First Law of Progress in Theoretical Physics is that "You will get nowhere by churning equations," in "Why the Renormalization Group is a Good Thing," in A. Guth, K. Huang, and R. Jaffe, eds., *Asymptotic Realms of Physics* (Cambridge, Mass.: MIT Press, 1983), p. 1; and then he shows that sometimes this is not so. Similarly, relabeling should not matter, but of course in actual practice it really does. (Here, the relabelings give us presumed good degrees of freedom of the weak interaction rather than of the strong interaction.) Good labels give us good handles onto the world – and that matters to us enormously.

30. This is perhaps a strong claim – but I think correct, especially from the point of view of most practicing problem-oriented physicists. See N. D. Mermin, "Bringing home the atomic world." There is, as well, a visible light analogy to the *K*-zero story, including regeneration, in terms of plane and circularly polarized light and suitable polarizers. See, for example, chapter 11 in vol. 3 of R. P. Feynman, R. B. Leighton, and M. Sands, *The Feynman Lectures on Physics: Quantum Mechanics* (Reading, Mass.: Addison-Wesley, 1965).

31. On classification and its peculiarities, see for example D. Hull, *Science as a Process* (Chicago: University of Chicago Press, 1988), a study of contemporary controversy in systematics. In the text I leave out the possibility of a continuum model of classification, largely because our everyday experience of species emphasizes discreteness.

32. For surprises that are more a matter of our mistaken theorizing (often because we do not appreciate the actual importance of certain mechanisms) see the wonderful R. Peierls, *Surprises in The-*

oretical Physics (Princeton, N.J.: Princeton University Press, 1979).

33. Only by running a simulation (an actual run-through) of the labels and rules in interaction might one see their systemic consequences. More generally, the surprises of the Invisible Hand and of the Cunning of Reason may be taken as providential. See J. Viner, chaps. 1 and 2 in *The Role of Providence in the Social Order* (Princeton, N.J.: Princeton University Press, 1972). In fact, physicists refer to some of these surprises as "spontaneous" (as in spontaneous symmetry breaking), marking their lack of homology. See chapter 4.

34. See Viner, chap. 2 in *Role of Providence,* on the providence that allows in the commerce between nations (because of their different abundances of goods) for the chance for brotherhood among men (and presumably women).

35. One turns on the rules by notionally altering the coupling constants, such as the electron's charge e (the coupling constant is actually $e^2/\hbar c$). Were the coupling constant zero, the exchange rules of electromagnetism would not be in force, and labels associated with it such as electrical charge would be of no significance.

A similar abstracting move is made by John Rawls, in *A Theory of Justice* (Cambridge, Mass.: Harvard University Press, 1970), when he speaks of "the original position." But physicists do not so readily concede the abstractness of the fundamental abstraction. This abstraction is actual Nature for them.

As for the Adam Smith story, the division of labor is presumably a consequence of a *tâtonnement* process combined with innovation that responds to price signals, so that a division of labor would provide for comparative advantage.

36. Technically, when physicists construct formal mathematical spaces, such as the four-dimensional spacetime of special relativity, namely a four-dimensional space with a particular metric, they are automatically providing for freedom and necessity, plenitude and nondegeneracy. The symmetries or classification rules of such a space (the equivalence of spacetime points, the so-called Poincaré transformation) say which labels are the same (degenerate) and how they are the same – here how space and time are the same. What is left over are the distinct possibilities, and they are the plenitude that will be fulfilled.

37. See chaps. 4 and 5 on "Sticky Systems" and on "Economy and Perversion," in Krieger, *Marginalism and Discontinuity,* especially chapter 5's discussion of fetishism, which grounds much of what I have said in this chapter.

4. THE VACUUM AND THE CREATION

1. As for fields, we might also have two approaches – such as taking up the slack in the work of particles or the story of path dependence – and again we might show that these are formally the same.

2. A. Pais, *"Subtle is the Lord"* . . . : *The Science and Life of Albert Einstein* (New York: Oxford University Press, 1982), p. vii.

3. Of course, Nature effectively hides other degrees of freedom from us, without much effort on our part. They may not appear in the specific heat, the temperature being too low and they are not excited, or they are degenerate and only appear when an external field is applied. Here, I am concerned especially with those degrees of freedom which set a stage or which we work to hide (rather than those that just happen to be too feeble to make their appearance felt in our situation, although that too is stage setting).

4. Here the degree of irrelevance is measured, respectively, by combinatorial considerations and entropy, the renormal-

ization group, and isospin symmetry and quantum mechanical considerations of energy fluctuations (or below-threshold behavior or virtual states – there is not enough energy to make them real and so lasting). Put differently, the thermal excitation (kT) is much less than the energy needed to push something out of hiding. See note 54 on preparing a system.

5. K. G. Wilson, "The Renormalization Group and Critical Phenomena," *Le Prix Nobel* (1982): 55–87, at p. 71 and p. 74.

6. S. Weinberg, "Why the Renormalization Group is a Good Thing," in A. Guth, K. Huang, and R. Jaffe, eds., *Asymptotic Realms of Physics* (Cambridge, Mass.: MIT Press, 1983), pp. 15–17.

7. P. W. Anderson, "Lectures on Amorphous Systems," in R. Balian, R. Maynard, and G. Toulouse, eds., *Ill Condensed Matter* (Amsterdam: North Holland, 1979), pp. 159–261, at p. 162.

8. Matter (and here I include radiation), as Somethings, is comparatively smoothly and regularly changing, even if quantized, for it is still homogeneous – and so it might even be added up as we discussed in chapter 1. But vacua are discontinuously connected (here literally by a falling down, reminding one of the biblical Fall). The vacua are incommensurable and not additive – rather, they succeed each other.

9. This also recalls the kabbalistic descriptions of the Creation as "the breaking of the vessels." In a different sense, the appearance of matter itself is ascribed to the violation and also breaking of a symmetry: CP violation, the baryon-lepton transition allowed in grand unified theories, and the disequilibrium that occurs within the Big Bang scenario.

10. These are very old questions, to be sure. See R. C. Dales, *Medieval Discussions of the Eternity of the World* (Leiden: Brill, 1990). For example, the notion of eternity is seen as an attack on Providence and the Creation, the traditional solutions to where something comes from.

There is, as well, an inverse problem – why there is nothing rather than something – when we consider the immanence of God everywhere. The modern equivalent is the very low energy-density of the physical vacuum – namely, the low value of the cosmological constant. See S. Weinberg, "The Cosmological Constant Problem," *Reviews of Modern Physics* 61 (1989): 1–23.

I have not emphasized that creation is an act of imaginative and textual and literary power, a theme that is important in studies of the rhetoric of science and Scripture.

11. E. Leach, *Genesis as Myth* (London: Cape, 1969).

12. I am not claiming that theater is a representation of everyday life, although it might well be. And it may be the case that any conception we have of our everyday life comes from projecting theater into it. What I am saying is that theater is a remarkable abstracted world, almost by fiat claiming what is to count and what not – at least if a theater works.

13. I speak here of "taming" because what is perhaps of greatest interest here is the mode of appearance of the degrees of freedom (for example, crystalline vibrations and phonons), rather than their number. Of course, degrees of freedom must be excited ("active") and independent of each other if we are to readily keep track of them. If the temperature is low enough, high-frequency (high-energy) degrees of freedom are not excited. And if a system is even mildly coupled, higher-order interactions (anharmonicities) link together what we took as independent stable degrees of freedom, often leading to shorter lifetimes for particles under such perturbing influences.

14. See chapter 1, note 13, which describes (à la P. W. Anderson) a variety of

discontinuous breaks, due to symmetry breaking for the most part. Note, as well, as we discuss later in this chapter, that one can usually find a way of, and a point of view for, making such discontinuous breaks smooth and incremental. But, in general, physicists are not smooth evolutionists about creation in the physical realm – but see Kenneth Wilson's work (note 42, below) in which "the search for analyticity" is an important theme.

15. See chapter 3 on "quite rarely." Also, gravitational forces are not canceled or shielded by particles with negative mass (no antigravity), and so the addition of the effects of many masses is quite potent. Electrical and magnetic forces are also "long range," meaning they drop off as inverse square powers of distance ($1/R^2$) rather than abruptly. Not only nearby molecules but also distant ones – especially if there are lots of them (say uniformly distributed out to infinity, so to speak) – can contribute to the force we feel. But their electrical force can be shielded or neutralized, and that is one contributing reason to why matter is stable. On the other hand, nuclear (strong) forces and weak forces are of short range, at least at the usual energy scales. At very high energies the relevant distances are very small, about a ten-thousandth the size of a nucleus, and then even these forces seem of longer range, with inverse square behavior.

16. It is fair to ask why the proton, and why within the nucleus? A physicist might warrantedly say it is only within such distance that the strong (or nuclear) force shows itself and so it might be compared, and particles such as electrons or neutrinos do not exhibit this force at all. Protons exhibit all the forces. Still, it is conventional that the proton is used as the test particle. In fact, if the distance between particles is about a ten-thousandth of the size of the nucleus, the forces are all roughly of the same strength, and the test particles might be less specific. This set of conventions is invoked when we are describing unification schemes rather than hierarchies.

17. If we have a large number of similar individuals interacting with each other, as in a crowd of people or in a gas or in a crystal filled with electrons, it may be quite difficult to develop any perturbative approach. One must find a good vacuum for the collection of individuals. But sometimes the particles (or Somethings) of this good vacuum are very close to the bare individuals themselves – as in a Fermi liquid's quasiparticles, where we find what P. W. Anderson, in chap. 3 of *Basic Notions of Condensed Matter Physics* (Reading, Mass.: Benjamin/Cummings, 1984), calls "adiabatic continuation" from the bare to the interacting state, free or valence electrons smoothly translating into quasi-electrons (at least for low-level excitations).

18. I have been imprecise here. What is crucial is the mass of the particle, for that determines the energy or temperature when copious pair creation can take place that does not lead to immediate annihilation. It is more or less true that stronger forces are represented by heavier particles, or at least larger differences in mass between particles in the same super-family.

19. A bit of technical metaphysics: Now why is there something, our matter-filled universe, rather than nothing, namely the "emptier" universe of just energy and fields? The answer physicists give is that the less symmetric matter-antimatter-filled world is, it turns out, more stable than is the empty universe (defined by very high fields or energy densities). When some matter-antimatter pairs appear as fluctuations in the emptiness, the more stable, ordered, less symmetric configuration grows, eventually

thermalizes, and then we have the cooling off subsequent to the Big Bang itself. Our current evidence suggests that the early universe was equally balanced between matter and antimatter. The dominance of what we call matter in our universe is a consequence of CP violation, baryon non-conservation, and disequilbrium in the cooling down process.

I do not address why we have our particular universe, a very different yet traditional question. See, for example, M. H. Krieger, "Temptations of Design: A Meditation on Practice," *Research in Philosophy and Technology* 10 (JAI Press, ed. F. Ferré, 1990): 217–30, reprinted as part of "The Manufacture of the Sacred, the Reenactment of Transcendence, and the Temptations of Design," in M. H. Krieger, *What's Wrong With Plastic Trees? Artifice and Authenticity in Design* (Westport, Conn.: Praeger, 2000), pp. 75–94. Currently popular physicists' answers range from the uniqueness of our doing physics and observing the universe to claims about the existence of many other universes, perhaps characterized by parameters rather different than ours. These are much like the range of the traditional answers.

20. If there is the possibility of supercooling and so metastability, this is a description of a freezing-like change, a first-order transition, to be triggered by random noise if we are below the transition temperature. If we are concerned just with broken symmetry (and so second-order transitions as well), then triggering is the spontaneous choice of direction we discussed earlier. Here "spontaneous" means that it is in effect without cause, or rather that we have no precise handle on its cause.

While I have in the text spoken in terms of cooling down a system and the appearance of orderly Somethings, physicists will of course also speak of heating up the system so as to liberate some frozen-in degrees of freedom (see section V). Only if it is hot enough, only if one is "above threshold," will such a degree of freedom be seen (as in the production of a particle). As the system gets even more energetic, that particle type will eventually become unstable since the ambient temperature allows for reactions that annihilate it. We shall want to cool down the system again if the particle is to be seen as stable.

21. I have, to be sure, "linearized" the story, ignoring some important detailed features of inflation and thermalization. The inflation scenario, in whatever form, is an answer to how the world could be so orderly, at all. See, for example, A. D. Linde, *Particle Physics and Inflationary Cosmology* (Char, Switzerland: Harwood, 1990), or *Inflation and Quantum Cosmology* (Boston: Academic Press, 1990).

22. Hegel speaks, in *The Phenomenology of Mind* (1807), of the structural identity of Geist, Wissenschaft, and Bildung.

23. As we shall discuss later, the great analytic device of historiography or of physics is to find an incremental or continuous path through the incommensurability of such discontinuous transitions; for example, what it might mean to become a crystal. See the preface and introduction to M. H. Krieger, *Marginalism and Discontinuity* (New York: Russell Sage Foundation, 1989). See also note 40.

24. In the quantum mechanical realm, there exist situations in which these excitations do not occur or occur quite improbably, since the Somethings are all finite in size and that size is large compared to the average thermal energy. Put differently, some degrees of freedom are not excited and so cannot contribute to the specific heat. In superconductivity, energy is required to break up a Cooper pair; this energy gap does not allow for small excitations, and hence there can be resistanceless flow. In the Mössbauer

effect, the probe's impact in the illumination of one atom is spread over the very large number of atoms in a macroscopic crystal, and hence lighting up the world can have minuscule side effects. Of course, in each case we still may see other Somethings.

25. And time, as sequentiality and succession, "begins" at Creation. But physicists measure time logarithmically – "10^{-43} seconds after the Big Bang" – so they never get to a beginning (log 0 equals minus infinity). See for example and for comparison, Augustine's discussion of time as narrative and of God's inaccessibility in this world and this time (the present), in bk. 11 of the *Confessions;* and Kant's of time as succession and hence narrative, again, in the *Critique of Pure Reason (Transcendental Aesthetic).* Book 11 of the *Confessions* proves to be a prooftext for physicists concerned with the early universe. See S. Weinberg's comment to this effect in "The Cosmological Constant Problem": "This quote [from Augustine] is not merely a display of useless erudition" (p. 15). See also John Freccero's discussion of time in Dante and its Augustinian origins in *Dante: The Poetics of Conversion* (Cambridge, Mass.: Harvard University Press, 1986).

26. In many actual phase transitions, we can see each side (of the transition temperature) but not the edge, where there are infinities (at least in principle and almost in practice, too). Physicists take the edge off, so to speak, by examining the transition in terms of variables through which it is not discontinuous or is at least less discontinuous.

27. On liminality, see V. Turner, *The Ritual Process* (Chicago: Aldine, 1969).

28. P. Ramond elegantly describes perturbation theory in just these terms, that of finding a good ground state so that the perturbations are simple and they are well separated from the ground state. P. Ramond, *Field Theory* (Reading, Mass.: Benjamin/Cummings, 1981), pp. 88–89. I quote the passage in Krieger, *Marginalism and Discontinuity,* p. 150.

29. Another such transition point is a threshold for particle production, as in the threshold for the production of charmed particles which then leads to a jump in the ratio of the production of hadrons to leptons in scattering.

30. Note that I take the ground state of a quantum system to be functionally equivalent, for my purposes here, to equilibrium.

See note 54 for a more precise definition of thermal equilibrium.

As a consequence of the deeper or justifying feature, one achieves an orderly vacuum by a means of invoking a constraint, and thus many degrees of freedom are tamed:

CONSTRAINT (to achieve an orderly vacuum)	DEEPER or JUSTIFYING FEATURE
Thermal-Equilibrium / Uniformity	Ergodicity
Rigid	Broken Symmetry
Repetition	Invariance
Alignment	Causal or Correlated
Smoothness	Continuum Limits
Gaussian/ Bell-Shaped	Statistical Independence
Insusceptibility	Optimality

My use of "causal" here will disturb those who try to distinguish it from "statistical" – but I doubt that many practicing physicists (especially those who employ correlation functions) will mind.

31. See Anderson, *Basic Notions,* pp. 49–51. Here I follow chap. 2 in Anderson. Note that liquid crystals are only finitely repetitive, and so rigidity as we under-

stand it every day does not follow in this case.

32. Anderson, *Basic Notions,* pp. 11ff., offers a host of reasons for why we might expect such broken symmetry.

33. While in the main these are the situations that concern physicists, there are as well many exceptions (especially at the forefront of research) which are eventually likely to be domesticated to these requirements – at least if the subject is to be physics (rather than engineering, chemistry, . . .). Of course, the requirements are modified by the domestication work they do, and so there is no static criterion of what is physics. And sometimes there is radical change over rather shorter periods.

34. On limits and continuum limits in physics, see chapter 1, note 47, on K. G. Wilson. See also M. H. Krieger, chap. 2 of *Marginalism and Discontinuity,* and "Theorems as Meaningful Cultural Artifacts," *Synthese* 88 (1991): 135–154.

35. The world still proves resistant to the central limit theorem, and so there need to be methods of taking into account outliers, namely, robust statistics such as the median and the trimmed mean. See Krieger, *Doing Mathematics,* chapter 2.

36. Several provisos: (1) Here, I have ignored saddle points and the like. Minimax stories then replace the optimality story. (2) Of course, thermal equilibrium is an optimum, but earlier in the text I took it statistically rather than thermodynamically. (3) Degrees of freedom are damped because at equilibrium we pick up just the normal modes of the system and they are now the good degrees of freedom. Anharmonicities, representing many more degrees of freedom, are in general comparatively small.

The Somethings defined by small disturbances of the orderliness of smoothness, bell-shapedness, and insusceptibility might be, for example: scale-symmetric structures as measures of roughness; autocorrelation; and springlike vibrations. On disturbances of independence, see Krieger, "Theorems as Meaningful."

37. For a much larger list, see the chart on pp. 67–68 of Anderson, *Basic Notions,* describing the consequences of broken symmetry.

38. In such a phenomenology as I have provided here, the emphasis is on analytic description rather than definition. One recognizes this strategy by the circularity of the analysis, there being no defined terms that are supposed to ground the exposition. The phenomenology justifies itself by its capacity to make us see what we already know in new ways. More conventional analysis might then begin.

39. Richard Feynman's explanations, in terms of interfering phases, for how certain principles actually work, such as the principle of least action, are notable for their rarity in the literature. See, for example, R. P. Feynman, R. B. Leighton, and M. Sands, chap. 19, in vol. 2 of *The Feynman Lectures on Physics* (Reading, Mass.: Addison-Wesley, 1964); and R. P. Feynman, chapter 2 of *QED* (Princeton, N.J.: Princeton University Press, 1985). Similar accounts for methods of steepest descent are given by E. Schroedinger, *Statistical Thermodyamics* (Cambridge: Cambridge University Press, 1964), pp. 31–33.

40. See A. J. Haymet, "Freezing," *Science* 236 (29 May 1987): 1076–80. According to Haymet, this idea is due to Kirkwood. Here, the free energy is assumed (counter to intuition) to be a smooth or analytic function of the density, a density whose amplitude of regular spatial modulation is just what implies that we have a crystalline order. That amplitude of modulation is called an order parameter. Given this analyticity assumption, as the temperature declines, the order parameter becomes larger (namely, nonzero, so indicating there is crystalline order). Note

that at lower temperatures than the transition point, excitations or the articulations of the order parameter (Somethings, vibrations of the crystal lattice) decrease with temperature. Put differently, as random thermal motions decrease (with decreased temperature) the system as a whole is more orderly, and there is less energy available to excite the system's more energetic orderly states.

Note that this example is for a first-order transition, while the next is for a transition of second order. Also note that we might want to give an account of freezing, say, in terms of what happens at the liquid-solid interface, in part a molecularly incremental process.

41. Moreover, concerning discontinuity, an account of the discontinuous creation of the vacuum in terms of interacting molecules does so by saying that discontinuity and the incommensurability of symmetries both appear "in the infinite volume limit" (of infinite numbers of particles yet constant density, for example). In such limits seemingly continuous functions, such as the free energy, become nonanalytic at the transition point. The infinite slope (or sharp change in the slope) of the graph of the order parameter at the transition point or critical temperature similarly depends on being in the infinite volume limit – most particularly because only then might one speak strictly of a broken symmetry or of a rigid system as in a spontaneously magnetized bar of iron.

For second-order phase transitions, all of this is strictly true; above the transition point, the orderliness (of spontaneous magnetization, for example) does not appear in bulk matter. Still, as you approach the transition point from above, say, there are larger yet microscopic regions that are coherent, having a net magnetization. Moreover, for first-order transitions where metastability is possible, there could be an incremental disorderliness; for example, below the transition point, where there is a fluctuation that is a small region of liquid ice (that is, water). See note 40 for details.

42. For example, physicists believe it is possible to find some point of view, some parametrization (even the dimensionality of space, as in Wilson and Fisher's $4-\varepsilon$ expansion) through which a discontinuous transition appears smooth.

43. More precisely, in a first-order transition, the two kinds of vacua actually coexist at the transition temperature (ice and water), and the higher symmetry (water) can persist alone metastably at lower temperatures in the case of supercooling; while in a second-order transition they exist together in the trivial sense that the higher symmetry is present at the transition temperature only to be lowered immediately (to greater orderliness) once that temperature is decreased at all.

44. Put differently, the ground state has a unique symmetry. See also note 47. For these physicists, the Great Chain of Being would seem to have to have some breaks or kinks, so to speak.

45. Henry Adams, "The Rule of Phase Applied to History," in E. Stevenson, ed., *A Henry Adams Reader* (Garden City, N.Y.: Anchor, 1958).

46. Energy refers to Helmholtz free energy, for example; volume refers to action and variational principles.

47. For some systems (called "frustrated"), such as antiferromagnets and spin glasses, there seem to be many different vacua, and the physicist's problem becomes one of redefining the vacuum so there is only one (for example, the vacuum is an ensemble of such ground states). The detailed vacuum we encounter may depend on the details of the external conditions, how we approach that vacuum. Good experimental setups try to take advantage of this dependence. Note that

in general ground-state degeneracies are to be "split" by external fields that violate the previous symmetry. I have ignored quantum mechanical zero-point energies here.

48. On Dirac's analogy to the rare gases and to X rays (which can be the consequence of missing inner electrons or "holes" in the otherwise filled shell structure), see H. Kragh, *Dirac* (Cambridge: Cambridge University Press, 1990), pp. 95–96. Removing an electron from the sea, a gap or hole in this sea turns out to be a positive energy, positively charged anti-electron (or a "positron"); while the electron that once occupied that hole now sits atop the sea, and is just a positive energy everyday electron – the pair of particles now making up a Something.

Haymet, "Freezing," claims that the crystalline structure is predictable by testing out the mathematically conceivable possibilities and looking for the lowest free energy.

49. See Krieger, chapter 2 of *Marginalism and Discontinuity,* on phase integrals and on integration by parts for such justifications. Another version of this is offered by a mathematician: "Banach was fond of saying that mathematics is the study of analogies between analogies; one may optimistically draw from his aphorism that in mathematics, every apparent coincidence points to an underlying cause which awaits discovery." J. P. S. Kung, M. R. Murty, and G.-C. Rota, "On the Redei Zeta Function," *Journal of Number Theory* 12 (1980): 421–36, at p. 421. Or, to quote from Schroedinger (*Statistical Thermodynamics,* pp. 36–37): "One of the fascinating features of statistical thermodynamics is that quantities and functions, introduced primarily as mathematical devices, almost invariably acquire a fundamental physical meaning. We had examples in the Lagrangian parameter μ [in doing the Stirling's approximation deduction of the canonical ensemble], the maximum z [from the Darwin-Fowler method], and the sum-over-states or partition function."

50. Note that we are referring to three vacua here – the ground state of the solid, the particle physicists' vacuum, and the ground state of a particle. Fluctuations here are due to the absolute temperature and Planck's constant both not being zero.

51. Namely, $e^{-E/kT}$ is either close to one or close to zero. Here I am also thinking of the Fermi level of a degenerate gas and its excitations.

52. Such rules are always expressed as a ratio or difference of the two energies: for example, $e^{-\Delta E/kT}$ or $1/(\Delta E - i\Gamma)$, where kT is a thermal energy and Γ is a measure of "width" or energy uncertainty, for thermodynamics and quantum mechanics respectively.

53. As for the Heisenberg principle, the space and time average charge or field might well be zero; still, there are positive and negative objects transiently present (to be sure, balancing each other). Hence, physicists speak of the creation and annihilation of virtual (rather than real) electron-positron pairs.

As for the second law, generally dissipations, such as a specific heat, are proportional to fluctuations. And if there were no classical fluctuations, then the specific heat would be zero; and that is only possible at absolute zero of temperature. See L. D. Landau and E. Lifshitz, pt. 1, secs. 20–21, and chap. 12 of *Statistical Physics,* 3rd ed. (Oxford: Pergamon, 1980); and note 54 below. That the specific heat is strictly positive is "equivalent" to the stability of matter and the second law of thermodynamics.

Also, the relative probability of a fluctuation is just e^S, where S is the entropy of that state, and near equilibrium, S (fluctuation of size x) = S (equilibrium)

$+ x^2/2\ \partial^2 S/\partial x^2$, so the probability of a fluctuation is a Gaussian or bell-shaped curve ($\partial^2 S/\partial x^2 < 0$ for entropy maximization).

54. This is the equipartition theorem. Also, Number of States for the System ≈ (Total Energy/k)$^{\text{Degrees of Freedom}}$. Entropy $= k \log N \approx kD \log E/k$, and $1/T = \partial(\text{Entropy})/\partial E \approx kD/E$. Namely, $kT \approx E/D$ = average-energy per degree-of-freedom, as expected (actually it is $kT/2$), where k is Boltzmann's constant. See chapter 3, note 18.

"If a system is very weakly coupled to a heat bath at a given 'temperature', if the coupling is indefinite or not known precisely, if the coupling has been on for a long time, and if all the 'fast' things have happened and all the 'slow' things not, the system is said to be in thermal equilibrium" (R. P. Feynman, *Statistical Mechanics* [Reading, Mass.: Benjamin/Cummings, 1972], pp. 1–6, at p. 1).

55. In the quantum field theoretic realm, we would say that the time scale we are concerned with is not too small (see Feynman's remark in the previous note), so energy uncertainty and effective temperature is also small, and so fluctuations are in general not large compared to the energy needed to create (activate) new particles. See Anderson, *Basic Notions*, pp. 70–72. See also chapter 1, note 30.

56. This is, strictly speaking, a story of second-order phase transitions.

57. See note 28.

5. HANDLES, PROBES, AND TOOLS

1. So-called phenomenological theories in physics are distinctive not because they are analogical, but rather because they explicitly eschew deeper or more microscopic explanations and mechanisms, and they relate measured phenomena to more generic models and their properties.

2. The practical historical problem is to describe which divisions have worked well in the past, which of those are appropriate now, and how they need to be modified for this situation.

3. Of course, some of the time the resistance is expressed more totally, in the inability of the conceptual structure to incorporate many puzzling features.

4. In Hacking's terminology, it has been a story of representation and not of intervention. I. Hacking, *Representing and Intervening* (New York: Cambridge University Press, 1983).

5. See S. Traweek, *Beamtimes and Lifetimes* (Cambridge, Mass.: Harvard University Press, 1988). See also M. H. Krieger, *Marginalism and Discontinuity: Tools for the Crafts of Knowledge and Decision* (New York: Russell Sage Foundation, 1989), pp. xvii–xix; and chapter 7, and references therein, on tools, toolkits, and guild and craft.

6. Obviously I have in mind a Kantian theme from the *Critique of Pure Reason:* What must the world be like so that we can have knowledge of it?

7. Even in the quantum world, where measurement does disturb the system, there is a well-prescribed way of taking that world as objective, having its own wave function more or less independent of us. Put differently, quantum mechanics may be taken to say "It ain't nothing till I call it" (as an umpire might say); but conventional interpretations of that saying do not alter the effective objectivity of the world. See the discussion of John Wheeler in J. Bernstein, *Quantum Profiles* (Princeton, N.J.: Princeton University Press, 1991). See also note 24.

8. An organismic biologist might use entirely different models, say an animal, and different modes of getting at the world, such as dissection or ecological observation. It is not clear that "probing" is the best way of describing such work, although it is not hard to conceive of the biologist's work in such terms.

9. See P. Galison, "History, Philosophy, and the Central Metaphor," *Science in Context* 2 (1988): 197–212.

10. In time, in the matter of course, that assimilation has a reciprocal effect and alters our everyday pictures. The latter is the province of the literary and intellectual historian, who might show how Newtonian ideas affected poetry and everyday notions of the cosmos, for example. See, for example, M. H. Nicolson, *The Breaking of the Circle* (New York: Columbia University Press, 1960).

11. These are all the ways we can get hold of a black hole.

12. See, for example, P. W. Anderson, chap. 3 in *Basic Notions of Condensed Matter Physics* (Menlo Park, Calif.: Benjamin/Cummings, 1984), on the principle of "adiabatic continuity." The quasiparticles that make up a Fermi liquid are incrementally "dressed-up" electrons (hence adiabatic continuity), clothed by the crystal lattice environment (and conduction band) in which they reside, still in many ways very much like their naked origins (the valence electron of an atom).

13. If there are chemical reactions or phase transitions, or the unfreezing of new degrees of freedom (as in particle production thresholds), the specific heat will reflect these separately.

At high enough temperatures, the electronic vibrations or transitions of the metal become prominent; and we have a blackbody, "composed" of photons. Then, heating up a metal bar changes it from red-hot to white-hot, the temperature change indicated by a change in the distribution of photon occupation numbers, the ones of greater energy and hence "whiter" now dominating.

14. Moreover, seemingly different handles, ones that are supposed to be equivalent in principle, will be the right degrees of freedom for different contexts. Do we want to treat a field quasi-classically, say in terms of electric and magnetic fields (or in terms of a vector potential), or quantum mechanically in terms of annihilation and creation operators?

Counting up many degrees of freedom (in order to estimate the specific heat, say) in large comparatively unconstrained systems such as a gas in equilibrium or a heated metal bar (a blackbody) critically depends on getting the active degrees of freedom right. Are the electronic modes of excitation to be counted, and when and how? The story of Planck's blackbody law is the story of figuring out which were the degrees of freedom that ought to be counted, and when and why. It is a continuation of the story of Maxwell and Boltzmann ($e^{-E/kT}$), but now with a new and crucial twist, quantized units of energy, hν, where h is Planck's constant and ν is frequency. But see T. S. Kuhn, *Black-Body Theory and the Quantum Discontinuity* (New York: Oxford, 1978), for its more technical origins. See also note 18.

15. Similarly, a rigid body made of many molecules, by the fact of its rigidity has only a very small number of degrees of freedom (namely, its position and orientation and their momenta), much smaller than do all of its molecules as such. Vibration of that body, treating it as elastic, increases the displayed degrees of freedom dramatically. Now if that rigid body were a crystalline structure, those vibrations are nicely tamed by the symmetries of the crystal. Conversely, in a very different example: After the Big Bang we know which particles, which degrees of freedom, can appear at which temperatures – they fall out of thermal equilibrium as the universe cools down.

16. More global handles – such as averages, measures of flow rate, and geometrical similarity – are the right degrees of freedom here. I am thinking, respectively, of "statistical continuum limits" (K. G. Wilson, "The Renormalization

Group: Critical Phenomena and the Kondo Problem," *Reviews of Modern Physics* 47 [October 1975]: 773–840), Reynold's numbers, and scaling variables. See chapter 1, note 47.

17. Hesse would call this a formal analogy. See M. Hesse, "Models and Analogy in Science," in vol. 5 of the *Encyclopedia of Philosophy* (New York: Macmillan, 1967), pp. 354–55.

18. From Einstein and Planck we learned that: (1) the energy of the quantum was proportional to frequency, so that energy was quantized in these units; and (2) those frequencies might be thought of as the vibration frequencies of springs or oscillators, while the square of the size of the oscillation would be proportional to the number of such quantized oscillators. The amplitudes we discuss in chapter 3 might be seen to represent the particles exchanged that make for a transition, their effective number and mass proportional, respectively, to what we have here called size-squared and frequency (which is complex for virtual particles).

Referring to note 13, we have here shown how physicists can take that red-hot piece of material as being made up of many little springs or oscillators. More technically, the complex set of vibrations and interactions of the atoms in the material may be characterized by the frequencies and amplitudes of the light they emit and absorb. See note 30 below.

19. Of course, an inverse square central force will not be linear and will give us a different kind of ground state and its excitations, as in the hydrogen atom.

For great sensitivity, physicists might want to work at an unstable point. Say all we are interested in is some null hypothesis and so we are sensitively looking for no effect. Small probings will have very large effects. But it is then hard to be sure just what is the source of any large effect we do see, whether it be noise or an outside probe.

20. An analogy is made to the best-known force – gravity and its inverse square force ($\approx 1/r^2$); $r = 0$ corresponds to $t = 0$, namely $T = T_c$; and if $r = 0$, then the gravitational force is infinite, while if $t = 0$ then l is infinite. (Of course, as we approach these infinities the models' assumptions may well break down. More adequate models then give finite results.)

21. For physicists, the difference between what are called phenomenological theories and more fundamental ones lies not in their being analogical or not, but rather in the degree of microscopic detail they engage in. See note 1.

22. C. D. Goodman, "Spin-isospin Sound: Going at the Nucleus with Tongs versus Hammer," *Bulletin of the American Physical Society* 28 (April 1983): 725–26.

23. And to change an occupation number is equivalent to tapping a spring.

24. Again, even in quantum mechanics, where a probe's effect is unavoidably finite, we develop a notion of the gentle probe and so can say what it means for a probe not to be present. I am not disagreeing with P. A. M. Dirac, when he says that "we can observe an object only by letting it interact with some outside influence. . . . There is a limit to the fineness of our powers of observation and the smallness of the accompanying disturbance" (*Principles of Quantum Mechanics* [Oxford: Clarendon, 1958], pp. 3–4). Rather, we may develop a notion both of probing and of what is already there yet unprobed that allows us to acknowledge what Dirac says and to fulfill our everyday notions of objectivity, albeit modified, as Mermin shows. See N. D. Mermin, "Bringing Home the Atomic World; Quantum Mysteries for Anybody," *American Journal of Physics* 49 (1981): 940–43; and *Boojums All the Way*

Through (New York: Cambridge University Press, 1989), pp. 81–94. We both impute and conceive of the state vector or wave function that defined that state before we do the measurement, a state vector that presumably predicts what we shall then see. So, jets of particles or ejecta from an explosion are to be explained by a notion of what there was before the interaction or explosion. All of this is not a philosophical claim, but rather one about the everyday practice and beliefs of actual professional physicists. See also note 7.

25. Technical definitions of probing, in terms of response functions and scattering theories, try to capture what it is we are doing when we are actually probing. Such generic technical definitions are modified as physicists invent new modes of probing that produce nice physics.

26. Actually stable equilibrium, and here thermal equilibrium. See Feynman's remark in chapter 1.

27. Mathematically, we might say that we divide the world into analytic parts and pole terms. And, again, there may he ways, topologically, of smoothing these out, so that a pole or a phase transition point is on a rather more smooth manifold, our intersection of that manifold (our setup, in the world or laboratory) being the accidental rather than the essential source of discontinuity.

28. They are called, among other names, bifurcations, symmetry breaking, chaos, percolation. For examples, see chapter 1, note 13.

29. Good probes couple minimally to a field, just enough and in enough ways to be sensitive yet have reproducible responses (which might not happen if they were too sensitive or too varied in their ways of coupling) – and then the field had better appear to be what physicists would call well behaved: smooth, charge conserving, and so forth. As physicists discover new general features of a field, "minimally" is redefined so that we now probe features that once were ignored.

30. M. Kac, "Can One Hear the Shape of A Drum," in K. Baclawski and M. D. Donsker, eds., *Probability, Number Theory, and Statistical Physics: Selected Papers* (Cambridge, Mass.: MIT Press, 1979), pp. 474–96.

31. See, on rigidity, P. W. Anderson, *Basic Notions of Condensed Matter Physics* (Menlo Park, Calif.: Benjamin/Cummings, 1984), pp. 50–51. It is often pointed out, for example either by Anderson or by R. Feynman, in *The Character of Physical Law* (Cambridge, Mass.: MIT Press, 1965), that the size and rigidity of apparatus, in contrast to the small scale and nonrigidity of much of the quantum world, makes for the probabilistic nature of measurement of the quantum world.

32. On the other hand, and in a rather different range of sensitivity to the acts of the observer, the anthropic principle would argue that the world depends on our appreciation of it. See J. Barrow and F. Tipler, *The Anthropic Cosmological Principle* (Oxford: Clarendon, 1986). Most radically, the observer is in effect part of the world's initial conditions. Of course, that we can conceive of the system as if no probe were present does not mean that measurement will not surely disturb the system. See note 7.

33. For example, if, in doing a series of measurements in an atomic beam apparatus, we see five lines bunched up together, we might reasonably infer that the spin of the atom was 2. For each atom, the "outcome" of which line it ends up on (its spin's z-projection) is otherwise unfixed (and meaningless) unless it is measured. See also N. D. Mermin, "Bringing Home the Atomic World; Quantum Mysteries for Anybody."

34. The history of beta radioactivity (electron-emitting nuclei) is rife with concern with how an electron could "fit within" a nucleus.

Or, we have the notion of virtual particles, whose presence is felt through their small perturbative effects or in their being liberated (made real, released) if the temperature or energy density is high enough.

35. Here I am thinking of the liquid-drop and statistical models of the nucleus. Now, insofar as a shield is permeable, there will be greater inelasticity; for more stuff that is inside can almost get out, and so it may well actually escape with help from a probe. Such shielded objects may be hard to get hold of by a single good degree of freedom; sharp-enough probes may stimulate such systems to fall apart.

More technically, this is the lesson of the fluctuation-dissipation theorem of Einstein, modified by Onsager. The dissipation or resistance or inelasticity a probe encounters is of exactly the same origin as the spontaneous fluctuations (here, within the permeable shield) of what is being probed. See D. Chandler, chap. 8 on the Onsager Regression Hypothesis, in *Introduction to Modern Statistical Mechanics* (New York: Oxford University Press, 1987).

Permeability means there will be comparatively many "can almost get out" fluctuations. Now, those fluctuations occur along the good degrees of freedom of an object, and a probe is most nicely influential through the same (internal – yet soon to be external) degrees of freedom. But that means that in probing there will be penetration or deformation of the shield or skin – permeability leads to potential inelasticity.

36. The latter is called inclusive scattering. On deep inelastic inclusive scattering, see, for example, J. D. Bjorken, "Feynman and Partons," *Physics Today* (February 1989): 56–59.

37. Correspondingly, on the theoretical side, the limiting processes and continuum limits which give us derivatives and velocities in particle motion, and which give "statistical continuum limits" (Wilson, "The Renormalization Group") in studies of fields and of condensed matter, provide unique results "no matter how" we take the limits. The good degrees of freedom are fetishes. Phenomenologically, a fluid has just a pressure, a magnet has just a magnetization, although the fluid or the magnet is composed of fluctuating microscopic atomic configurations that are quite varied and complex. In the limit of macroscopic scales, some microscopic details of probing must become irrelevant. See chapter 1, notes 40 and 47. On fetishism, see Krieger, chap. 5 on "Economy and Perversion," in *Marginalism and Discontinuity.*

38. On Paley, see N. C. Gillespie, "Divine Design and the Industrial Revolution: William Paley's Abortive Reform of Natural Theology," *Isis* 81 (1990): 214–29.

39. None of this is to be taken as a prescription, philosophical and formalist though it may sound. It is what physicists do.

40. Much of this section comes from part of Krieger, "The Physicist's Toolkit," in chap. 7 of *Marginalism and Discontinuity,* or from the original article with the same title in *American Journal of Physics* 55 (1987): 1033–38. More details, especially for those philosophically inclined, are to be found there.

41. H. L. Anderson, ed., *Physics Vade Mecum* (New York: American Institute of Physics, 1981).

42. See F. Yates, *The Art of Memory* (Chicago: University of Chicago Press, 1966). Much of what I understand about all of this derives from some discussions with W. J. Ong, S.J., almost forty years ago in Palo Alto.

43. One might even argue that what is so interesting about lattice gauge theories is that such fields (which were, historically, to be freed of their æthers) now have in effect a crystalline ether in which they survive (at least until the lattice spacing is set to zero). More generally, since Dirac, the vacuum has become a lively place once again. See chapter 4, especially note 48.

44. This resolution is expressed, for example, by the final states in the scattering matrix; trivially, by the existence of constants of motion as in action-angle variables; and by stable quasiparticles representing a sum of most interactions.

45. P. Levi and T. Regge, *Dialogo* (Princeton, N.J.: Princeton University Press, 1989), p. 48.

46. Again, "Long-lived theoretical entities, which don't end up being manipulated, commonly turn out to have been wonderful mistakes" (Ian Hacking, *Representing and Intervening,* p. 275). See also the quote from Maxwell in the Preface.

47. In the *Philosophical Investigations,* by his choice of example, Wittgenstein puts it nicely: "Bring me a slab" (§20). See also J. Berger, *Ways of Seeing* (Harmondsworth: Penguin, 1972), on possession by sight.

48. Presumably, I am describing much of what T. S. Kuhn, in *Structure of Scientific Revolutions* (Chicago: University of Chicago Press, 1970), has called normal science; and many of the models I offer are the paradigms and exemplars through which practice is enacted. See also the preface. Finally, following Hacking, *Representing and Intervening,* my main concern in chapters 1 through 4 has been to describe the modes of representing the world, and in 5 the modes of intervening in that world. What is interesting is how modes of representing build in the possibility of intervening, and how modes of intervention presume upon representation.

49. See G. C. Rota, "Fundierung as a Logical Concept," *Monist* 72 (1989): 70–77, on "letting." S. S. Schweber has written how American and pragmatic is this conception. See, for example, "The Empiricist Temper Regnant: Theoretical Physics in the United States, 1920–1950," *Historical Studies in the Physical Sciences* 17 (1986): 55–98. On the breaking down of a tool, see M. Heidegger, secs. 15 and 18 of *Being and Time* (New York: Harper and Row, 1962).

50. For example, field theories provide a technical grammar of probing, saying just how to probe and learn about the world. And a theory might account for all fundamental interactions in terms of the exchange of particles, real or virtual. For example, electromagnetic theory involves exchanges of photons or charged particles. See Barrow and Tipler, *Anthropic Cosmological Principle,* for a study of the knowability of Nature and its implications for the form of theory and of Nature.

51. Again, knowledge of this sort might be taken as a gift of our electromagnetic heritage.

6. PRODUCTION MACHINERY

1. Galileo Galilei, "From *The Assayer,*" in *The Essential Galileo,* ed. and trans. M. Finocchiaro (Indianapolis: Hackett, 2008), p. 183. While this passage is widely quoted, its context is worth noting. Galileo is arguing against authority and for independent mindedness (which, here, is thinking mathematically).

2. S. Soames, *Philosophical Analysis in the Twentieth Century* (Princeton, N.J.: Princeton University Press, 2003); R. Sokolowski, *Introduction to Phenomenology* (Cambridge: Cambridge University Press, 2000).

3. E. H. Lieb and J. L. Lebowitz, "The Constitution of Matter: Existence of Thermodynamics for Systems Composed

of Electrons and Nuclei," *Annals of Mathematics* 9 (1972): 316–98.

4. Technically, one asks, what is the "infinite volume limit"?

5. The mathematical limiting process may not be "uniformly convergent" and then the limit of the properties might well not be a smooth function. Think here of a rounded corner, whose radius approaches zero.

6. It is much more difficult to discern the fact that Ising matter close to the phase transition point looks the same at all scales by using the counting framework developed by Onsager, and it is rather apparent in a counting method that counts at various scales rather than one by one, or methods that do not count up at all but "innocently" build in that scaling almost from the beginning based on the algebra of the devices (matrices) used to do the counting.

7. E. H. Lieb, "The Stability of Matter: From Atoms to Stars," *Bulletin of the American Mathematical Society* 22 (1990): 1–49, at pp. 1, 27.

8. It should be kept in mind that the Dyson-Lenard result had a proportionality constant (which we would want to be as small as possible) of about 10^{14} (Rydbergs/hydrogen atom), and that the Thomas-Fermi proof got that down to about 5.6!, while the more mathematical inequalities only achieve 7.3, while the expected "semi-classical" result suggests 4.9. See E. Lieb and R. Seiringer, *The Stability of Matter in Quantum Mechanics* (Cambridge: Cambridge University Press, 2009).

One might hope that the crucial features revealed in the sequence of proofs might be read back into the original Dyson-Lenard proof. It would seem that Dyson doubts that the Thomas-Fermi proof's features could be read into his and Lenard's papers.

9. M. H. Krieger, *Constitutions of Matter* (Chicago: University of Chicago Press, 1996) and *Doing Mathematics* (Singapore: World Scientific, 2003) discuss this example extensively.

10. J.-M. Maillard has given talks where he says that the 2-d Ising model is nothing other than elliptic curves.

11. See chapter 5 of Krieger, *Doing Mathematics*. André Weil is the source for the "Rosetta Stone."

7. AN EPITOME

1. Again, we might call those analogies the Kantian categories and transcendental conditions for doing physics.

Index

MARTIN H. KRIEGER, who was trained as a physicist at Columbia University, has been a Fellow at the Center for Advanced Study in the Behavioral Sciences and at the National Humanities Center. He is author of *Marginalism and Discontinuity: Tools for the Crafts of Knowledge and Decision* (1989), *Constitutions of Matter: Mathematically Modeling the Most Everyday of Physical Phenomena* (1996), and *Doing Mathematics: Convention, Subject, Calculation, Analogy* (2003). He is on the faculty of the University of Southern California, and has taught at Berkeley, Minnesota, MIT, and Michigan. He is a Fellow of the American Physical Society.